SPECTROGRAPH DESIGN FUNDAMENTALS

In recent years enormous changes have occurred in the field of optical spectrometry. The classical spectrometer has become obsolete and the spectrograph, in combination with the CCD detector, now offers a vastly superior approach. Although the basic optical principles remain unchanged the design considerations are very different, and in many cases more demanding. However, developments in computer ray-tracing and computer-aided design have coped with these extra impositions and have allowed the construction of a new generation of spectrographs.

The book covers the general principles of spectrographic design, and the practical and engineering aspects of a broad range of spectrographs and spectrometers. This allows the reader to make an informed choice of instrument. It will be of particular use when none of the immense array of manufactured spectrographic and spectrometric instruments is suitable for a specialised task. The book deals with materials and methods of construction and includes suggestions for the choice of optical table, the design of slit mechanisms, and adjustable mirror, grating and lens mounts, with suggestions for the alignment and calibration of the finished instrument.

Spectrograph Design Fundamentals describes the design and construction of optical spectrographs. It will be a valuable resource for academic researchers, graduate students and professionals in the fields of optics, spectroscopy and optical engineering.

JOHN JAMES is an Honorary Senior Research Fellow at the University of Glasgow and a Fellow of the Royal Astronomical Society. He is the author of *Student's Guide to Fourier Transforms*, also published by Cambridge University Press, now in its second edition.

SPECTROGRAPH DESIGN FUNDAMENTALS

J. F. JAMES
Honorary Research Fellow
University of Glasgow

CAMBRIDGE UNIVERSITY PRESS
Cambridge, New York, Melbourne, Madrid, Cape Town, Singapore, São Paulo

Cambridge University Press
The Edinburgh Building, Cambridge CB2 2RU, UK

Published in the United States of America by Cambridge University Press, New York

www.cambridge.org
Information on this title: www.cambridge.org/9780521864633

© J. F. James 2007

This publication is in copyright. Subject to statutory exception
and to the provisions of relevant collective licensing agreements,
no reproduction of any part may take place without
the written permission of Cambridge University Press.

First published 2007

Printed in the United Kingdom at the University Press, Cambridge

A catalogue record for this publication is available from the British Library

ISBN-13 978-0-521-86463-3 hardback
ISBN-10 0-521-86463-1 hardback

Cambridge University Press has no responsibility for the persistence or accuracy of URLs for
external or third-party internet websites referred to in this publication, and does not guarantee that
any content on such websites is, or will remain, accurate or appropriate.

Contents

Preface		*page* ix
Acknowledgements		xi
1	**A brief history of spectroscopy**	1
2	**The relevant regions of the electromagnetic spectrum**	6
	2.1 The limits of optical spectrography	8
3	**Geometrical optics**	10
	3.1 Rays and wavefronts	10
	3.2 Instrumental optics	11
	3.3 Centred systems	12
	3.4 Gaussian optics	12
	3.5 Optical layout	20
	3.6 Apertures, stops, fields, irises and pupils	20
	3.7 Ray bundles	23
	3.8 The Helmholtz–Lagrange invariants	23
	3.9 Surface brightness	25
	3.10 Black body radiation	25
4	**Optical aberrations**	28
	4.1 The Seidel aberrations	29
	4.2 Zero-order aberration	29
	4.3 First-order aberrations	29
	4.4 Theorems	35
	4.5 Aberration coefficients for mirrors	37
	4.6 The achromatic doublet	39
5	**Fourier transforms: a brief revision**	41
	5.1 Fourier transforms	41
	5.2 Theorems	42
	5.3 Convolutions	42
	5.4 The Wiener–Khinchine theorem	45

	5.5	Useful functions	45
	5.6	More theorems	49
	5.7	Aliasing	51

6 Physical optics and diffraction — 52
- 6.1 Fraunhofer diffraction — 52
- 6.2 Two-dimensional apertures and oblique incidence — 55

7 The prism spectrograph — 57
- 7.1 Introduction — 57
- 7.2 The traditional prism spectrograph — 57
- 7.3 The focal curve theorem — 59
- 7.4 The Littrow mounting — 59
- 7.5 The Pellin–Broca prism — 60
- 7.6 Focal isolation — 61

8 The plane grating spectrograph — 63
- 8.1 The shape of a monochromatic line spectrum — 63
- 8.2 Blazing of gratings — 67
- 8.3 Apodising — 68
- 8.4 Order overlap and free spectral range — 69
- 8.5 Grating ghosts and periodic errors — 70
- 8.6 The complete grating equation — 72
- 8.7 Differential dispersion — 76
- 8.8 Mounting configurations — 76

9 The concave grating spectrograph — 89
- 9.1 The Rowland grating — 89
- 9.2 The concave grating as a spectrograph — 92
- 9.3 The concave grating as a monochromator — 96
- 9.4 The aberrations of the Rowland grating — 97
- 9.5 Practical details of design — 98

10 The interference spectrograph — 101
- 10.1 The phase angle — 101
- 10.2 The Fabry–Perot étalon spectrograph — 102
- 10.3 Fabry–Perot theory — 102
- 10.4 The Fabry–Perot monochromator — 104
- 10.5 The Fabry–Perot CCD spectrograph — 108
- 10.6 Fore-optics — 110
- 10.7 Reference fringes — 111
- 10.8 Extraction of the spectrum — 111
- 10.9 Choice of the resolution and gap — 112
- 10.10 The 'crossed' Fabry–Perot spectrograph — 113

11 The multiplex spectrometer — 114
- 11.1 The principles of Fourier spectrometry — 114
- 11.2 The multiplex advantage — 117

12 Detectors — 120
- 12.1 Silver halide photography — 121
- 12.2 Elementary electronic detectors — 122
- 12.3 Detectors with spatial resolution — 123
- 12.4 Exposure limitations — 124
- 12.5 CCD software — 125
- 12.6 CCD calibration — 126
- 12.7 Spectrograph calibration — 126

13 Auxiliary optics — 128
- 13.1 Fore-optics — 128
- 13.2 The astronomical telescope as fore-optics — 130
- 13.3 Focal reducers — 132
- 13.4 Schmidt-camera spectrography — 134
- 13.5 Scattered light and baffling — 134
- 13.6 Absorption cells — 136
- 13.7 Fibre optical input — 137

14 Optical design — 139
- 14.1 First steps — 139
- 14.2 Initial layout — 139
- 14.3 The drawing board — 140
- 14.4 Computer ray tracing — 140
- 14.5 Refinement of the optical design — 141
- 14.6 Requirements of a ray-tracing program — 144

15 Mechanical design and construction — 150
- 15.1 The optical layout — 150
- 15.2 Optical materials — 162
- 15.3 Transparent optical materials — 163
- 15.4 Reflectors — 163
- 15.5 Metals for construction — 165
- 15.6 Other materials — 167

16 Calibration — 168
- 16.1 Sensitivity calibration — 168
- 16.2 Wavelength calibration — 169
- 16.3 Small spectral shifts and radial-velocity measurement — 170
- 16.4 Absorption measurements — 170

17 The alignment of a spectrograph 172
   17.1   The optical alignment 172
   17.2   The focus 173
Appendix 1   Optical aberrations 175
Appendix 2   Wavelengths of spectral lines for calibration 179
Appendix 3   The evolution of a Fabry–Perot interference spectrograph 183
Appendix 4   The common calibration curve in silver halide spectrophotometry 186
   Bibliography 187
   Index 188

Preface

Thirty-eight years ago, together with my colleague the late Robert Sternberg, I wrote a book entitled *The Design of Optical Spectrometers*. It described the state of the art as it was at that time after the great advances which had come in the previous ten years, and it was intended for people who wished to build a spectrometer tailored to a specific purpose, where perhaps one of the commercial designs was inadequate, unsuitable, unnecessarily cumbersome, or expensive.

When at last the time came to consider a new edition it became clear that the technology had changed so much that the classical optical spectrometer, in the sense of *monochromator*, was more or less obsolete and that later developments such as the desktop computer and the charge-coupled device had restored the spectrograph to its former eminence. The restoration in no way annulled the optical improvements of the previous 30 years but new constraints posed new problems in design. These problems are now solved and the solutions are presented here.

The fundamentals of optical design have not changed, but the constraints are now all different, and such properties as flat fields are needed where before they could be largely ignored; and focal ratios matter again when previously we could design everything so that such trivia as spherical aberration and coma could be neglected.

There is, as always, a gap to be bridged between the elegant theory presented in the undergraduate textbooks and the practical spectroscopic instrument standing on the laboratory bench or bolted to the Cassegrain focus of an astronomical telescope. The gap is partly in the limitations imposed by the curse of non-linearity in geometrical optics and the contumacious aberrations it produces, and partly in the sometimes obstinate physical properties of the materials of construction. As always in scientific instrument making, the art is in knowing what must be precise and exact and what can be left go hang at a crude level. There are tricks in this trade just as in any other. The traditional engineer's description of a physicist is 'someone who designs a box that must be screwed together from the inside'. (There is a parallel physicist's definition of an engineer: someone who, when asked 'what is three times four?' will get out a slide rule and answer 'about twelve'.) There is an element of truth in this and some of the hints in these pages may help to refute the calumny.

This work is thus intended for the new generation of researchers who desire high efficiency in an instrument tailored to their own particular purpose, and who have access to a mechanical workshop of moderate size where an instrument of their design can be constructed. There has been no attempt to venture into the new fields of optical resonant scattering spectroscopy, tunable laser spectroscopy or other specialised techniques and so the book is directed to chemists, astronomers and aeronomers as much as to physicists.

Manufactured spectrographs are in no way deficient, but are necessarily compromises, both in performance and cost, and are often intended for teaching or dedicated to a routine task such as sample analysis. A specially designed instrument has no difficulty in excelling them for a particular investigation, particularly in academic research fields where new areas are being explored and where no established technology yet exists.

It is to the basic optical design and construction of such individual instruments that this book is dedicated. I also take the opportunity of acknowledging my late good friends and colleagues, Dr H. J. J. Braddick, Rob Sternberg and Larry Mertz, from whom I learned so much.

Acknowledgements

Many individuals and institutions have contributed to the research funds and facilities which yielded the knowledge and experience written down here. Chief among these are the following:

The Royal Society Paul Instrument Fund for numerous development grants

Royal Society Research grants for field-work support

The Royal Society and Royal Astronomical Society Joint Permanent Eclipse Committee for expedition funds

The Japan Society for the Promotion of Science for field-work support

The British Council, in Spain and Japan, for travel grants for collaborative development and field-work support

The Science and Engineering Research Council, for development grants, expedition funds and field-work support

El Instituto Astrofisica de Andalucia, for collaboration, hospitality and facilities during field-work

The Royal Greenwich Observatory, Herstmonceux, for test facilities and the use of telescopes while testing prototype instruments

l'Observatoire de Haute Provence for observing facilities and the use of telescopes for observation

The University of Texas McDonald Observatory for technical facilities during field-work

The 'Atenisi Institute, Nuku'alofa, Tonga, for facilities and hospitality during field-work

The University of Manchester Schuster Physical Laboratory workshop and drawing office staff for meticulous design detail and expert workmanship on many instruments

The University of Glasgow for an Honorary Research Fellowship while this book was being written

1
A brief history of spectroscopy

The earliest reference to optical spectroscopy that we have in modern times appears to be the *phenomenon of colours* in Isaac Newton's *Opticks*, in which he describes his famous experiments with prisms and the shaft of sunlight coming through the hole in his window shutter. There was much philosophical conjecture at the time but scientific silence from then on until William Hyde Wollaston (1766–1828) in 1802, also in Cambridge, used a lens to focus images of a narrow, sunlight-illuminated slit through a prism on to a screen. Wollaston appears to have observed the dark lines across the spectrum transverse to the dispersion direction but ascribed them to the divisions between the colours. He may be forgiven for this, because with a single lens the spectral resolution would have been derisory. At about the same time William Herschel (1738–1822) discovered the infra-red radiation by the rise in temperature of the bulb of a thermometer when he held it beyond the red part of the spectrum in his spectroscope. Joseph von Fraunhofer (1787–1826) saw more dark lines but did not guess or deduce their origin. The currently accepted explanation – the absorption of continuous white light by vapours in the atmosphere of the Sun – was given by Gustav Kirchhoff and Robert Bunsen in the University of Heidelberg who, we may be reasonably certain, passed a collimated beam through their prism before focusing it, and thereby secured a reasonable resolution. It is probably to these two but perhaps to Fraunhofer as well that we owe the classical form of the optical spectroscope: the sequence of collimator, prism and telescope. Little has changed fundamentally since then.

Developments in geometrical optics kept pace with developments in photography. From Lord Rayleigh we have the complete theory of the prism spectroscope and of the diffraction grating. Comparatively crude but useful diffraction gratings could be ruled on the bed of any screw-cutting lathe and many individual workers made gratings for their own use, but without passing on the art. Practical details of grating construction were given by R. W. Wood of Johns Hopkins University. At this stage though, they were little more than scientific curiosities emphasising the wave nature of light. This was at a time when the corpuscular theory was by no means dead.

However, diffraction gratings were first realised at a serious and practical level by H. A. Rowland, also of Johns Hopkins University, who first devised ruling engines capable of the necessary precision.

The prism spectroscope continued in its role as the chief instrument of astronomical botany – the taxonomy or classification of stars, that is – but the resolution available then was insufficient to do more than measure wavelengths, identify and discover elements and to allow wise comments on the mysterious 'fluted' spectra sometimes found in flames as well as in cool stars, so described by their resemblance, at low dispersion, to the flutings on the columns of classical temples. It was a major achievement for a spectroscope to be able to resolve the six angstroms separating the 'D' lines of sodium, and Rayleigh computed that a prism base length of 'at least 1 centimetre' would be needed for this, that is to say a resolving power of ~ 1000.

Sir Arthur Schuster achieved a resolving power of more than 10 000 at Manchester University with a twelve-prism[1] spectroscope made for him by Cooke of York, with 30 glass–air interfaces (excessive even in a modern coated zoom lens!) and prism bases of 25.4 mm. With incidences near the Brewster angle there should have been near perfect transmission for one state of polarisation, but in practice a transmission coefficient of about 0.05 made it virtually useless except as a sort of tour de force of the spectroscope-maker's art. The way forward in those days was the manufacture of ever-larger prisms from ever-denser glasses.

The finely ruled diffraction grating by contrast improved the spectral resolution immeasurably and led to the evolution of the spectrograph at the hands of H. Ebert, who in 1889 removed chromatic aberration and a great deal more besides by replacing the lenses of the collimator and the camera with a single concave mirror. The symmetry was intended to eliminate coma as well as chromatic aberration but the apparent symmetry is removed by the tilt of the grating and, although reduced, the coma is still there.

Ebert's invention was inexplicably lost until rediscovered by W. G. Fastie 50 years later, by which time the single mirror had been replaced by two separate mirrors in the now standard Čzerny–Turner configuration.

It was the invention of efficient vacuum pumps which led to the investigation of gas-discharge phenomena by Crookes and others and the consequent exploration of gaseous emission line spectra. This led in turn to the empirical discoveries by Balmer, Paschen and Rydberg which gave rise to the need for an explanation and the consequent invention of the Bohr model of the atom and the subsequent evolution of quantum mechanics.

---

[1] In practice six prisms each passed twice.

In the twentieth century the spectroscope developed in accordance with the best detector technology available at the time. Little happened before 1950 in the optical part of the spectrum although infra-red detectors made considerable progress, and at wavelengths below 0.8 μm the photographic emulsion was the universal detector, the spectroscopic art advancing along with improvements in its sensitivity and panchromaticity. By *c.* 1950 the whole electromagnetic spectrum from soft X-rays to about 1.2 μm was available to silver halide emulsions and the short-wave limit to practical spectrography was set by the reflection coefficients of the optical components, which were – and are – dismal below about 1200 Å. This limitation was offset by the gradual development of improved sources of XUV light such as the synchrotron for absorption spectrography and later spectrometry.

Semiconductor detectors improved the accessibility of the infra-red, and the photomultiplier, which became generally available *c.* 1950, revolutionised optical spectroscopy. This device yielded a detective quantum efficiency (DQE) in the region of 0.2–0.35 as opposed to the DQE of a silver halide crystal of about 0.0005 and, just as importantly, gave a linear response.[2] One of the bugbears of the silver halide emulsion, especially at the low surface brightness of many astronomical sources, was *reciprocity failure*, the inability to respond adequately to long exposures at a low light level. Consequently a monochromator with a photomultiplier detector, free from the constraints of low focal ratio, could easily out-perform a spectrograph and moreover could be made to concentrate its efforts on the parts of the spectrum important to the investigation being made.

Geometrical optical design was then brought to bear, resulting not only in high efficiency but in much greater resolving power, sometimes reaching 80–85% of the theoretical level imposed by the constraints of physical optics.

In parallel with the development of the grating spectrograph, interference spectrometry extended the high-resolution end of the art. Rayleigh again observed that the signal from a Michelson interferometer was the Fourier transform of the spectrum of the incoming light but little note was taken of it at the time except as a scientific curiosity, and it was Charles Fabry and Alfred Pérot in the University of Marseille who in 1899 devised the interferometer which revealed fine and hyperfine structure in spectral lines. Other simultaneous devices, such as the echelon grating,[3] the Lummer plate[4] and the echelle[5] gave equally high resolution but were far less convenient to use and were not widely copied. In the visible and UV region it is the

---

[2] Based upon the experimental finding that ∼1000 photoelectrons must be liberated within a microcrystal of silver bromide, irrespective of its size, to make it developable.
[3] A. A. Michelson, *Astrophys. J.*, **8** (1898), 37.
[4] O. Lummer, *Verh. Deutsch. Phys. Ges.*, **3** (1901), 85.
[5] G. R. Harrison, *J. Opt. Soc. Amer.*, **39** (1949), 522.

Fabry–Perot interferometer in its various guises which still rules the high-efficiency end of the spectrographic art.

Meanwhile high resolution was being achieved by more-or-less heroic means, with plane-grating instruments of two or three metres focal length and focal ratios of F/20 or F/30. This ensured at least that the optical resolution would be diffraction limited and that photographic exposures would take an inordinately long time. Academic research was simultaneously being carried out using concave gratings of up to ten metres radius, in spectrographs which filled a whole room in a laboratory or the whole of the Coudé focus room of a large astronomical telescope. There appears to have been some confusion at the time between *dispersion* and resolution, as though one depended on the other, and dispersion itself was taken as a measure of quality or performance. With the passage of time has come improved knowledge and a spectrograph much bigger than one metre or perhaps two in focal length is now a rarity. Provided that an adequate design is chosen, all the information that can be extracted from a spectrograph can be revealed with a 1.5-metre instrument.

The period between 1950 and 1970 saw a great surge in activity in the art of optical spectroscopy. This followed Jacquinot's revelation that 'L-R product' or 'efficiency' in a spectrometer, defined as the product of 'throughput' or 'light-grasp'[6] and resolving power, was a constant and that one quality could be traded for the other. There was consequently a huge blossoming of ingenious, weird and wonderful inventions, including the mock-interferometer,[7] the Sisam,[8] the Girard grille,[9] the Möbius-band interferometer,[10] the tilting Michelson interferometer,[11] the lamellar grating interferometer[12] and various 'chirped' and field-compensated Michelson Fourier spectrometers.[13] Much time and effort were expended to bring these ideas to practical use and the two famous 'Bellevue' conferences of 1957 and 1967 brought all these people together for an exchange of ideas.

However, this insubstantial pageant has faded and very few of the devices described there have survived the test of time–convenience–cost-benefit analysis.

The concept of L-R product is important, indeed fundamental. The idea of resolving power, defined as $R = \lambda/\delta\lambda$ where λ is the wavelength being observed and $\delta\lambda$ the smallest wavelength separation that can be distinguished as two separate

---

[6] A felicitous phrase from H. A. Gebbie.
[7] L. Mertz, N. O. Young & J. Armitage, *Optical Instruments and Techniques*, K. J. Habell, Ed. (London: Chapman & Hall, 1962).
[8] Spectromètre Interférentiel à Sélection par l'Amplitude de Modulation. P. Connes, *J. Phys. Rad.*, **19** (1958), 197.
[9] A. Girard, *Opt. Acta*, **7** (1960), 81.
[10] W. H. Steel, *Opt. Acta*, **11** (1964), 211.
[11] R. S. Sternberg & J. F. James, *J. Sci. Instrum.*, **41** (1964), 225.
[12] J. D. Strong & G. A. Vanasse, *J. Phys. Rad.*, **19** (1958), 192.
[13] L. Mertz, *J. Opt. Soc. Am.*, **49** (1959), iv; P. Bouchareine & P. Connes, *J. Phys.*, **24** (1963), 134; Y. P. Elsworth, J. F. James & R. S. Sternberg, *J. Phys. E*, **7** (1974), 813.

spectrum lines, is an old, dimensionless quantity and a useful parameter of quality in a spectroscope. 'Light grasp' on the other hand, also known as 'luminosité' or 'étendue' and generally denoted by the letter L, is less familiar and is a measure of the rate at which an optical instrument can process the power it receives from the incident electromagnetic field. It can be defined as the product of detector sensitive area and the solid angle subtended by the detector at the exit pupil of the system[14] or alternatively as the product of the area of the exit pupil and the solid angle subtended by the detector sensitive area. Both of these definitions are derived from the *Helmholtz–Lagrange* invariants of the system. The constancy of the product can be envisaged easily by referring to a simple prism spectroscope. If the entry-slit width is doubled, twice as much light passes through but the resolution is halved.

The concept is much more universal than this and it became apparent that high-resolution devices such as the Fabry–Perot interferometer, possessing a much higher L-R product than a grating spectrometer, could be used at moderate resolution to examine low-luminosity sources. Soon afterwards the 'multiplex' or Fellgett[15] (so named after its discoverer) advantage made Fourier-transform spectroscopy with a Michelson interferometer a highly desirable goal despite its formidable technological problems – now happily solved.

For more than 30 years the monochromator in one form or another ruled supreme until it was rudely overthrown by the arrival of the charge-coupled device, the 'CCD'. When cooled to the temperature of liquid nitrogen the CCD has a detective quantum efficiency of ~ 0.5 or more and it is, in effect, a photographic plate with an ISO rating in the region of 3 000 000 and a resolution of anything up to 100 line-pairs/mm.[16] The other improvements which it brought are legion: the lack of reciprocity failure, the enormous dynamic range in the region of 50 000,[17] the virtually instant read-out on to a computer screen, and so on. To develop its full potential in spectrography it has been necessary to return to geometric optics in spectrograph design, as there is a strict requirement for a flat field, and a distinct advantage in a low focal ratio; and there are modern advances which would have benefited the Victorian spectrograph had the problem been tackled at the time.

---

[14] Which for F-numbers ≥ 1.5 is related to the focal ratio F by $\Omega = \pi/4F^2$.
[15] P. Fellgett, *J. de Phys.*, **19** (1958), 187, 237.
[16] As compared with the fastest and consequently coarsest silver halide emulsions with ISO ratings in the region of 2000 and ~ 50 line-pairs/mm.
[17] Silver halide has a dynamic range of about 100: it saturates at ~ 100 times the minimum exposure needed to produce blackening.

2
The relevant regions of the electromagnetic spectrum

The extent of the electromagnetic spectrum is too well known to require description here. We shall be chiefly concerned with the so-called 'optical' region, of which the boundaries are determined partly by the methods of detection and partly by the methods of dispersing and analysing the radiation. What is common to all parts of the region is the type of optical element and materials of construction of spectroscopic instruments. Beyond the region on the long-wave side, coherent detectors, paraboloidal aerials, waveguides and dipole arrays are used, and on the short-wave (X-ray) side, optical elements other than diffracting crystals and grazing incidence reflectors are generally unknown. Radiation of 100 Å wavelength liberates photoelectrons with more than 100 eV of energy and the appropriate detection methods are those of radiography and nuclear physics.

Broadly we can identify six wavelength divisions appropriate to optical design techniques:

- 50–15 µm. The far infra-red (FIR), where bolometric, superconductor and semiconductor detectors are the chief methods of detection and measurement, and only specialised materials such as selenium, thallium bromide and various polymer resins such as sulphones have the necessary transparency to make useful refracting components. Optical elements may well be polymer plastics of high dielectric constant. Reflecting elements are coated with gold. New FIR-transparent materials with desirable optical properties are appearing all the time. Nevertheless, the principles of design which make use of them are unchanging and it is important to pay attention to their refractive index and partial dispersion or v-value when deciding whether to use them.
- 15–1.2 µm, 0.1–1 eV. The near infra-red, where there are many suitable refracting materials, and where semiconductor, superconductor and bolometric detectors prevail. This region includes the two 'windows', the 2–4 µm band and the 8–13 µm band where the atmosphere is transparent. The opacity does not much concern laboratory work, but is important if aeronomy, telecommunications or astronomical research is in view. The edges

- 1.2–0.35 μm, 12 000–3500 Å, 1–3.5 eV. The near infra-red, visible and near ultra-violet (UVOIR)[1] region where both semiconductors, photoconductors, photochemical reactions and photoemitters can be used, the latter directly as so-called 'photon-counters' (in reality photoelectron counters). This, not unnaturally, is the best understood region of the spectrum and where the greatest variety of refracting materials are known. It is also the region between 1 eV and 3 eV of energy, where interesting things happen in chemistry and physics, where atoms and molecules display their properties, where chemical reactions take place. At wavelengths above 8000 Å atmospheric transmission is patchy with extensive regions of opacity caused by water-vapour and oxygen molecules. Laboratory work is unaffected but aeronomers, astronomers and surveyors may have problems when they need long path lengths through the atmosphere.
- 0.35–0.20 μm, 3500–2000 Å, 3.5–6 eV. The far ultra-violet, where optical components must be selected carefully for adequate transmission[2] and phosphors are required to convert the radiation into longer wavelengths for registration by optical detectors. Again laboratory work is unaffected but the ozone layer of the atmosphere at 45 km effectively absorbs all extra-terrestrial UV radiation below 2900 Å. Laboratory sources emitting these wavelengths (a quartz-envelope mercury vapour lamp for example, which emits the 2537 Å Hg $^1S_0$–$^3P_1$ resonance line) are harmful and must be shielded, together with any surfaces which may reflect that radiation into the surrounding laboratory. Such sources may also stimulate the production of ozone (O_3) which has a characteristic 'electric' smell and has well-known corrosive properties on some materials such as rubber.
- 0.2–0.12 μm, 2000–1200 Å, 6–10 eV. The vacuum ultra-violet, where atmospheric gases absorb the radiation (O_2 below 2424 Å and N_2 below 1270 Å), where most materials have lost their transparency and where the reflecting powers of metals are beginning to decline. Optical instruments for this region must be in a vacuum chamber or in a non-absorbing atmosphere such as helium, which is transparent down to 911 Å, and other rare gases similarly down to their respective ionisation potentials. It is an interesting and not yet thoroughly explored region where autoionisation series spectra can be detected with their concomitant 'superallowed' transitions and broad line absorption spectra. Filling the spectrograph with helium does not ease the vacuum problem so far as leak prevention is concerned but it may ease the stress on thin-film windows.
- 0.12–0.01 μm, 1200–100 Å, 10–100 eV. The far- or extreme-vacuum ultra-violet, sometimes called the XUV, where radiation is ionising, where there are no transparent materials except in the form of thin (∼1–2 μm) films, where reflection coefficients are down to ∼0.1–0.2 and apart from synchrotron radiation there are no reliable, steady sources of radiation for absorption spectroscopy.

---

[1] **U**ltra-**V**iolet-**O**ptical-**I**nfra-**R**ed.
[2] Most optical glasses are opaque below 3500 Å.

Gas molecules have absorption cross-sections measured in megabarns (1 Mb = 10^{-18} cm^2) and this, in an optical path of $\sim$1 m, implies a need for a vacuum better than 10^{-4} mm Hg, which in turn requires a diffusion pump or, if especially pure samples are to be examined, a turbo-molecular pump. This is probably the most difficult region of the spectrum for which to design spectrographs and in which to do research. The vacuum requirements are demanding and not to be undertaken lightly.

At wavelengths below $\sim$100 Å materials start to become transparent again[3] and this is the beginning of the soft X-ray region where different techniques supervene.

In each of these regions of the spectrum the optical design requirements have much in common. There are exceptions.

- In the infra-red, alternative techniques such as Fourier-transform spectroscopy, usually with a Michelson interferometer,[4] are possible, are advantageous and are capable of enormous resolving power. A warning, however: the method lacks spectrometric simultaneity. A flickering source can have catastrophic effects on the spectrum unless rapid sawtooth scanning is used at a frequency well above the highest flicker frequency or below the lowest. In commercially available Fourier spectrometers this rapid scanning is unusual. Fourier spectroscopy has no advantage elsewhere than in the infra-red.
- In the UVOIR region the plane diffraction grating is the radiation analyser of choice although there are circumstances where a prism has advantages. The Fabry–Perot interference spectrograph, too, has its place where high resolution is needed or where very low light levels are encountered.
- In the vacuum ultra-violet between $\sim$2000 Å and 1200 Å prism and lens materials such as calcite and lithium fluoride are available with the required transmission but generally the concave grating possesses advantages which increase rapidly as the wavelength of interest decreases. At wavelengths below 2000 Å the concave grating is the analyser of choice. Below 1200 Å there is no realistic alternative.

So far as refracting materials are concerned for these various regions, technology is advancing all the time, especially in the infra-red and the Internet should be consulted for details of their transmitting and refracting properties and for the names of their manufacturers.

2.1 The limits of optical spectrography

An optical spectrograph or spectrometer of one sort or another can be made to give a resolution up to $R \sim 10^7$ without a struggle. It is difficult to imagine a practical application for a conventional spectrograph with a greater resolving power than

---

[3] Or the radiation more penetrating!
[4] But see Section 7.6.

this.[5] A Fabry–Perot étalon with a 3-metre spacer and a finesse of 50 will give $\sim 4.10^8$ at optical wavelengths, enough to reveal the natural widths of allowed atomic transitions, provided of course that you can first eliminate all the temperature effects. Extraordinary precautions are needed to ensure stability (in the source as well as the instrument) since the étendue is tiny and the exposure time long. It has the merit of being absolute, since the various quantities involved are directly measurable with a metre-stick. But at this sort of resolving power, more sophisticated spectroscopic techniques are available and preferable.

---

[5] The measurement of small Doppler shifts, for example, does not require high resolving power, only high spatial resolution at the focal plane. A wavelength shift of a small fraction of a resolved element is measurable.

3
Geometrical optics

This is the branch of optics which deals with image-forming instruments, including of course spectrographs and interferometers. Such instruments employ lenses, mirrors and prisms and it is the art of combining these elements to make useful devices which is the subject of the next two chapters.

3.1 Rays and wavefronts

Image-forming instruments are intended to project *images* of real objects on a *screen* or *focal surface* – usually though not always plane, and instrumental optics is the study of ways of doing this.

The subject comprises two chief parts: *instrument design* and *lens design*.

The former is the design of instruments so that light is conveyed through their optical components from one to another to arrive eventually at the focal surface. The technique is essentially a graphical one, using a drawing board or a computer drafting program to lay out the components at their proper places.

The latter involves the accurate tracing of rays through various optical surfaces of different types of glass, and reflections from mirrors of various shapes to achieve a correction of the optical aberrations and to ensure that light of all wavelengths from a point on an object is focused to a corresponding point on its image.

Traditionally this was done by *ray tracing*, the accurate computation of ray paths using seven-figure logarithms and trigonometrical tables, but now is done chiefly by iterative ray tracing in a small computer, using a program specifically designed for that purpose.

A word of warning: an optical system such as a spectrograph cannot be designed on a ray-tracing program. The proper positions for the various elements must be found by means of the drawing board or the CAD program. An iterative ray-tracing program can then be used to optimise the performance of the instrument by altering the positions of the optical elements, the curvatures and tilts of refracting and reflecting surfaces and their excentricities.

3.2 Instrumental optics

High-school teaching is that light is emitted from a luminous point, travels in straight lines outwards called 'rays', and that these rays can be caught by a lens and bent to form bundles which then converge to a point. If a screen is placed at the proper place, a point can be seen, the 'image' of the original luminous 'object'.

Light rays are a convenient fiction of course like the 'lines of force' in an electric or magnetic field, or the 'streamlines' of fluid mechanics. Following Huygens' principle, we postulate that light is emitted as wavefronts (which are also known as 'surfaces of constant phase') which spread outwards spherically from each point of an existing wavefront so that their envelope forms a new wavefront. They are slowed down when they pass through a refracting transparent medium such as glass or water. A spherically expanding wavefront from a point source can be converted by a properly shaped agglomeration of lenses to a spherical converging wavefront, which converges to form a point image of the point object. The 'point source' is another convenient fiction.

Other necessary concepts are:

- **Refractive index**   This is the ratio of the speed of light *in vacuo* divided by the speed of light in a refracting medium.[1] Refractive index is generally measured indirectly by observing the deviation of a ray as it passes through a prism for example.
- **Geometrical path**   This is the actual physical distance between a source of light and the place where the light falls on a screen and is observed. This is the distance which you measure with a ruler on a ray diagram on a drawing board.
- **Optical path**   This is the geometrical path multiplied by the local refractive index. Usually it is the sum of several sections where the light passes through different media, and if the refractive index, n, varies continuously it is the integral

$$\int_a^b n(s)\,ds$$

along the path of the ray.

Alternatively – and this is useful – it is the number of wavelengths of monochromatic light along the path, multiplied by the vacuum wavelength of the light.

The basic principle of image formation in an optical instrument is this: that the paths of the rays from a point on an object to its conjugate point on the image should all have the same optical path length.

---

[1] I use the word 'speed' here, since the speed, unlike the velocity, can be measured directly by ordinary physical experiments.

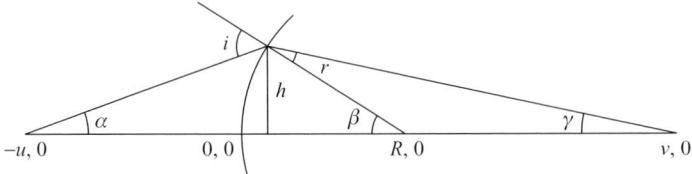

Figure 3.1 The basic diagram for ray tracing. It shows the various quantities involved in tracing a ray through a surface, using Snell's law to relate the angles i and r.

- **Reduced path** This is the geometrical path *divided* by the refractive index. For example, if you look into a swimming pool it appears to be shallower than it is. What you see – or would measure with a camera rangefinder for example – is the *reduced path* to the bottom of the pool.

All these are useful ideas in the field of optical design.

3.3 Centred systems

Optical image-forming instruments such as telescopes and microscopes generally have optical trains of spherical surfaces with the centres all lying on a straight line, the *optic axis* of the system. The use of spherical surfaces derives from the physical fact that these are the only surfaces which can be ground and polished with high accuracy with comparatively simple apparatus. One piece of glass ground against another with a suitable abrasive automatically produces a common spherical interface. To achieve other shapes, such as paraboloids, more elaborate methods are required but with contemporary technology are achieved at small extra cost.

3.4 Gaussian optics

The elementary theory of image formation is due to Carl Friedrich Gauss (1777–1855) and makes the assumption that all the angles between rays and the optic axis are small, so that the approximations $\sin\theta \simeq \theta$ and $\cos\theta \simeq 1$ are good.

By convention, optical diagrams are drawn with light entering the system from the left. Suppose (Fig. 3.1) that there are two transparent media of refractive indices n_1 and n_2, and a spherical interface between them of radius R with its *vertex* at the origin of a Cartesian coordinate system and the centre of the sphere at the point $C(R, 0)$. Suppose that a ray of light starts from a point on the optic axis $A(-u, 0)$ and reaches the interface at D where it is refracted according to Snell's law and continues in a new direction to meet the optic axis again at $B(v, 0)$. Draw the rays from C to D to B. Angles are measured clockwise from the optic axis to the ray.

3.4 Gaussian optics

Then,

$$h/u = -\alpha, \qquad h/v = \gamma, \qquad h/R = \beta,$$

and by ordinary geometry,

$$r = \beta - \gamma, \qquad i = \beta + \alpha,$$

and with Snell's law,

$$n_1 \left(\frac{1}{R} - \frac{h}{u} \right) = n_2 \left(\frac{1}{R} - \frac{h}{v} \right),$$

giving

$$\frac{n_1}{u} - \frac{n_2}{v} = \frac{n_1 - n_2}{R},$$

or, more usefully,

$$n_1 \left(\frac{1}{R} - \frac{1}{u} \right) = n_2 \left(\frac{1}{R} - \frac{1}{v} \right),$$

and this is a fundamental formula connecting the *object distance* $-u$, the *image distance* v, the *radius of curvature* R of the surface and the two refractive indices n_1 and n_2.

3.4.1 The thin lens

We next apply this result to two surfaces in succession, in other words, to a simple lens. The image distance after refraction by the first surface becomes the object distance for refraction at the second. The distance between the two vertices is taken as negligible. Then, with suffixes for the two radii and ticks for the distances from the second surface,

$$n_1 \left(\frac{1}{R_1} - \frac{1}{u} \right) = n_2 \left(\frac{1}{R_1} - \frac{1}{v} \right),$$
$$n_2 \left(\frac{1}{R_2} - \frac{1}{u'} \right) = n_1 \left(\frac{1}{R_2} - \frac{1}{v'} \right),$$

and u', the object distance for the second surface, equals v, the image distance for the first. Then, as before,

$$n_1 \left(\frac{1}{R_1} - \frac{1}{u} \right) = n_2 \left(\frac{1}{R_1} - \frac{1}{v} \right),$$

and for the second surface,

$$n_2\left(\frac{1}{R_2} - \frac{1}{v}\right) = n_1\left(\frac{1}{R_2} - \frac{1}{v'}\right).$$

Eliminating v between these and adding gives

$$\frac{n_1}{R_1} - \frac{n_1}{u} - \frac{n_2}{R_1} + \frac{n_2}{R_2} - \frac{n_1}{R_2} + \frac{n_1}{v'} = 0,$$

and putting, as usual for lenses in air, $n_2 = n$ and $n_1 = 1$, the result is

$$\frac{1}{v} - \frac{1}{u} = (n-1)\left(\frac{1}{R_1} - \frac{1}{R_2}\right).$$

An obvious and probably familiar corollary follows if we put the object distance equal to $-\infty$, for then

$$(n-1)\left(\frac{1}{R_1} - \frac{1}{R_2}\right) = \frac{1}{f},$$

where f is the *focal length* of the lens. The plane perpendicular to the optic axis where the image is formed is called the *focal plane* of the lens. There are obviously two focal planes, one either side of the lens itself, at distances f and $-f$ from the vertex. Where they cross the optic axis are the two *foci* or *focal points* of the lens.

The other familiar equation follows:

$$\frac{1}{v} - \frac{1}{u} = \frac{1}{f}.$$

This is the Gaussian lens equation.

NB Distances to the left (object side) of the lens are negative and those to the right (image side) are positive. This is the so-called 'sign convention'; one of many which plague geometrical optics, but the one which appears to work best and which is used throughout here. It is known generally as the 'Cartesian convention'. Another part of the convention is that surface curvatures are positive if the surface is convex to the incoming light, that is, if the centre of the sphere lies to the right of the vertex of the surface.

Notice that in general the optic axis can be defined as the line joining the two centres of the spheres which form the lens surfaces. In practice one tries to construct the instrument so that all the sphere centres lie on one line. This calls for some subtlety in manufacture and the accuracy with which a given lens element has its optic axis at the mechanical centre of its (circular) rim is called the *centration*.

Design note It is much more convenient when doing calculations to use *curvatures* and *powers* rather than radii and focal lengths. The Gaussian equation is

3.4 Gaussian optics

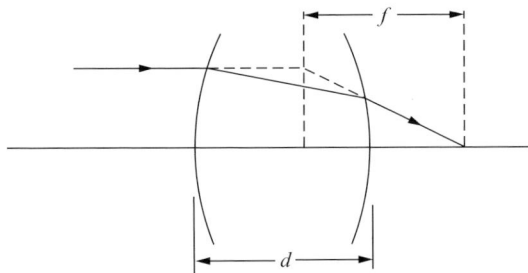

Figure 3.2 The geometry of a thick lens. The position of a virtual equivalent thin lens is one of the two *principal planes* which defines the position of refraction of a ray by any lens or system of lenses. The distance from this plane to the focal point is the focal length of the lens.

then
$$P = (n-1)(c_1 - c_2) = (n-1)\Delta c,$$
where $P = 1/f$ and is measured by opticians in reciprocal metres, or *dioptres*.

3.4.2 Thick lenses

We assumed above that the vertices of the two surfaces were coincident. In practice there is a distance d between them, where d is small compared with the object and image distances.

Repeating the argument, but with $u' = v - d$ and some tiresome algebra we find
$$\frac{1}{f} = (n-1)\left(\frac{1}{R_1} - \frac{1}{R_2} + \frac{(n-1)^2}{n}\frac{d}{R_1 R_2}\right).$$

This is sometimes called the 'lensmaker's equation'.

Confusion often arises when concave and convex mirrors are included. A certain way round the difficulty is to treat every optical component, lens or mirror, as if it were a lens, and to lay out the system on the drawing board as if it were a totally refracting one. Then a mirror concave to the left behaves like a lens with a positive focal length f and $f = R/2$ where R is the radius of the mirror. To be pedantic, the mirror can be treated as a lens of refractive index -1 and a flat first surface.

An important tip is to remember that the angle through which a ray is deviated on passage through a lens depends only on the height of incidence and the focal length.

3.4.3 Aplanatic surfaces and lenses

The aplanatic points U and U' of a surface are related by the fact that any ray through the surface directed towards U will arrive at U' after refraction. Their

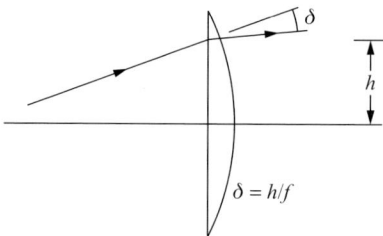

Figure 3.3 The refraction angle of a ray entering a lens depends *in the Gaussian approximation* only on the height of incidence and the focal length.

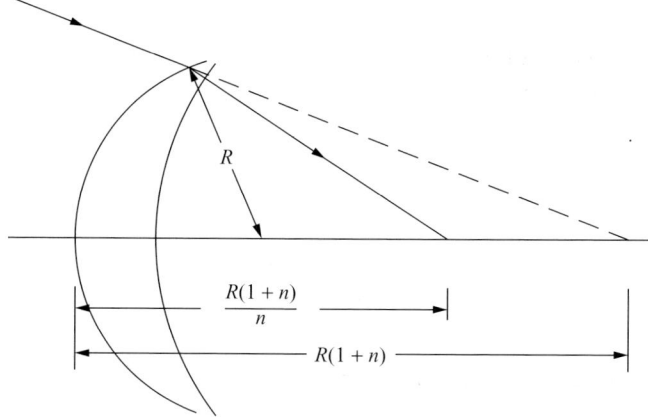

Figure 3.4 The converging aplanatic lens. The object is always virtual. Both object point and image point are at finite distances from the first surface vertex and are related to each other as in the diagram by the refractive index. The second surface position is not critical but it must be concentric with the paraxial image point. Perfect imaging holds for any incident ray directed initially towards the object point. The imaging is virtually aplanatic for a finite field surrounding the paraxial image point.

distances from the vertex of the surface are easily shown[2] to be related by

$$u = R(1+n) \qquad v = R(1+n)/n. \tag{3.1}$$

An image so formed in the vicinity of U' from a virtual object in the vicinity of U will thus be free from spherical aberration, coma and astigmatism.

A converging lens can be made with these properties provided the second surface is concentric with the image. The thickness of the lens is not of prime importance and such a lens is called an aplanat. In spectroscopy its chief use is to increase the

---

[2] A simple exercise for the reader, but failing this, see for example R. Kingslake, *Lens Design Fundamentals* (New York: Academic Press, 1978).

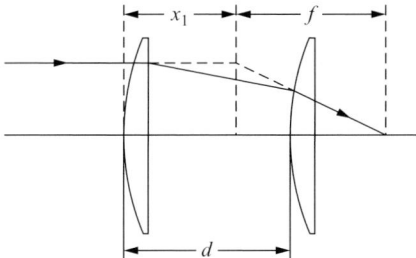

Figure 3.5 The geometry of a pair of thin lenses showing the position of a single equivalent thin lens at the appropriate principal plane.

convergence of a ray bundle so that it can be made to fall on to a smaller detector, thus reducing the detector noise which is area dependent.

3.4.4 Systems of two lenses

Consider two thin lenses of focal lengths f_1 and f_2 lying on the same axis and separated by a distance d (Fig. 3.5). It is usual to refer to the whole system as a 'lens' and to the two components as *elements*.

A ray coming from $-\infty$ parallel to the optic axis at a height h_1 above it suffers a deviation $\theta_1 = h_1/f_1$ at the first element and $\theta_2 = h_2/f_2$ at the second. This amounts to a total deviation $\theta_1 + \theta_2$, so that the focal length of the combination is $F = h_1/(\theta_1 + \theta_2)$.

With $h_2 = h_1 - \theta_1 d$ the h's and the θ's can be eliminated to give the focal length of the whole compound lens as

$$\frac{1}{F} = \frac{1}{f_1} + \frac{1}{f_2} - \frac{d}{f_1 f_2},$$

or, more usefully,

$$F = \left(\frac{f_1 f_2}{f_1 + f_2 - d} \right).$$

For practical purposes we need to know more than this. The usual question is 'where should one put a single thin lens, equivalent to the combination, which will produce an image at the same position and with the same focal length?' In other words, where should a single lens of focal length F be placed to secure an image at V?

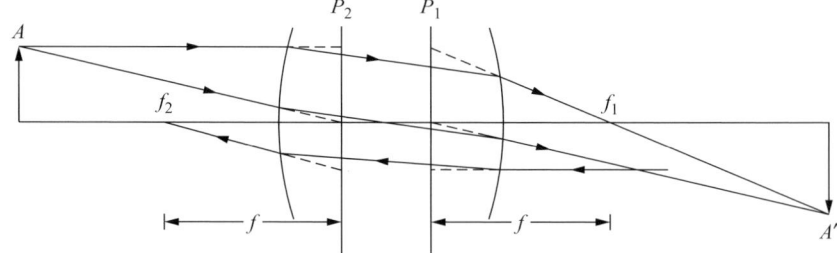

Figure 3.6 The principal planes and focal planes of a system of lenses. Only the first and last surfaces are shown and there may be any number of surfaces in between. Any such system of lenses possesses such a pair of principal planes which serve to determine the paths of rays traced through the system in either direction. Actual ray paths are drawn in full, construction paths are shown interrupted.

3.4.5 Principal planes

Suppose that the proper position is at a distance x_1 to the right of the first element. Then (Fig. 3.5)

$$d - x_1 = \frac{h_1 - h_2}{\theta_1 + \theta_2} = \frac{\theta_1 d}{\theta_1 + \theta_2} = \frac{Fd}{f_1},$$

so that

$$x_1 = d(1 - F/f_1) = \frac{d(f_1 - d)}{f_1 + f_2 - d},$$

and the simple equivalent lens must be placed at a distance x to the right of the first element to obtain the same-sized image of an object at $-\infty$ at the same place.

If the light were coming from the right:

- the focal length of the compound lens would be the same;
- the position of the simple equivalent lens would be

$$x_2 = \frac{d(f_2 - d)}{f_1 + f_2 - d}$$

to the left of the second element.

These two positions are called the principal planes of the compound lens. Where they cross the optic axis are the *principal points* of the lens.

It can be shown that any centred system of lenses and mirrors, however complicated, has two principal planes and two focal planes. They can be defined as follows (Fig. 3.6):

A ray, parallel to the optic axis, entering the lens system at P_2 emerges at P_1 at the same height above the axis and passes through the focus F_1.

Similarly a ray from the right entering the system at P_1 emerges at P_2 at the same height and passes through F_2.

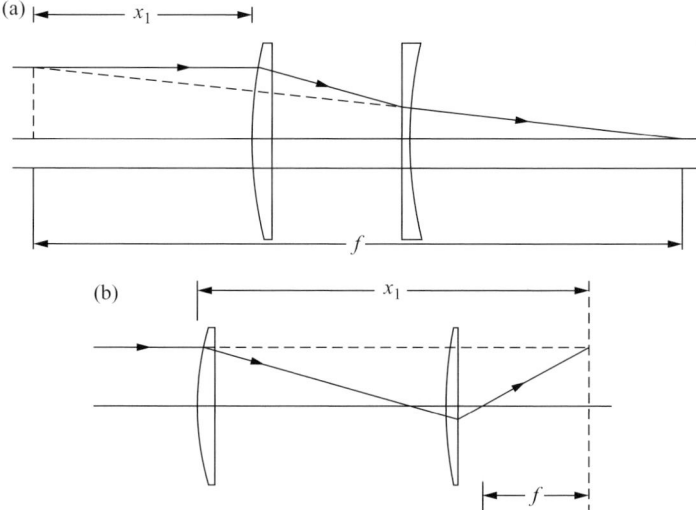

Figure 3.7 Two examples of lenses with principal planes outside the lens itself. (a) Represents a telephoto lens and (b) is a combination of positive lenses with a negative focal length, albeit giving a real erect image of a point at $-\infty$. The single equivalent lens would be at a distance x_1 to the right of the first vertex, and would be a negative lens giving a virtual image a distance f to the left of x_1.

3.4.6 Nodal planes

Two further points and planes can be defined, although for most purposes they are coincident with the principal planes. They are the *nodal planes*. There are several definitions, but a simple and practical one is as follows:

A ray from the left, crossing the optic axis at the nodal point n_2 at an angle θ to the optic axis, emerges from the other nodal point n_1 at the same angle. The nodal points coincide with the principal points unless the initial and final refractive indices are different. This circumstance might happen if a lens were designed to work with one end in water or oil for example.

For a single-element thick lens the principal planes are inside the glass at a distance $x_i = -f[(n-1)/n](d/R_i)$ from the surface i (where $i = 1$ or 2, obviously). In practice $x_i \simeq d/n$.

The principal planes are not necessarily between the elements nor even near the lens itself, as the examples in Fig. 3.7 show. A telescope in proper adjustment, for example, has no focal length and its principal planes are at $\pm\infty$.

The principal planes are also called the planes of unit magnification. If some other optical system to the left produces an image at the P_2 plane, the image will be transferred with the same size to the P_1 plane and will appear to be there to any subsequent system.

3.5 Optical layout

Once the positions of the focal planes and principal planes have been established, rays can be drawn from an object point, through the system, to the image point. The rules are:

- A ray from a point A on the object, parallel to the axis, entering at any height h meets the principal plane P_2 and emerges from P_1 at this same height and at an angle to make it pass through the focal point f_1.
- If a ray from the same object point, drawn to the nodal (or, in practice, principal) point of P_2 makes an angle ϕ with the optic axis it emerges from the nodal (principal) point of P_1 at the same angle ϕ and continues until it meets the other, previously traced ray, at a point A'.

Where they meet is the image of the object point A. The two points A and A' are said to be *conjugate*.

This construction is sufficient to establish the position of the image of any object and is generally the first step to be taken when designing a centred optical system.

3.6 Apertures, stops, fields, irises and pupils

3.6.1 The aperture

In a simple lens this is a self-evident concept. It can refer to the clear transmitting area of the lens or to the diameter if the lens has a circular rim. In a compound camera lens it is the diameter of the biggest parallel bundle of rays, all parallel to the optic axis, which is transmitted through to form an image. This may well be smaller than the aperture of the first element. This aperture may also be controlled by an *iris diaphragm* somewhere on the optic axis (though not necessarily inside the system), which then forms an *aperture stop*.

3.6.2 Numerical aperture

This is defined as $\mathrm{NA} = n \sin \theta_o$, where θ is the angle that the marginal ray makes with the optic axis, and is a measure of the intensity with which light can be focused on to an image. A similar and more familiar concept is the *F-number* or focal ratio, much used in describing camera lenses. The F-number of a lens is given by $F = 1/2 \tan \theta$ and for small values of θ is the reciprocal of twice the numerical aperture.

3.6 Apertures, stops, fields, irises and pupils

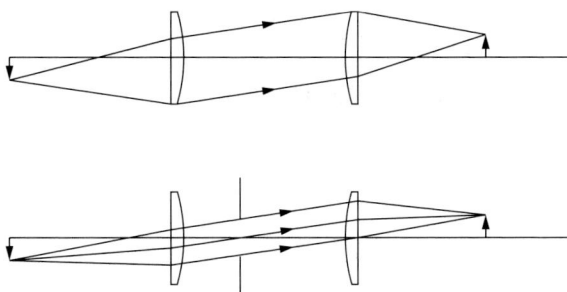

Figure 3.8 This shows the passage of a beam through a compound lens with air-spaced elements with and without an iris to control the ray bundles. The stop ensures a uniformly illuminated field and, as we shall see later, can be used to control the aberrations of the lens.

3.6.3 The stop

This is an opaque mask, generally perpendicular to the optic axis, in which there is a (usually) circular hole through which light can pass. Stops can be found anywhere in a centred system and may be a useful way of controlling the aberrations and hence the image quality that is produced. A telescope, for example, may well have a *field stop* placed at the position of an intermediate image – just in front of the eyepiece perhaps – where it reduces scattered light, increases the image contrast and produces a sharp edge to the field of view.

3.6.4 The field

In general the field of an imaging system is considered to be an angle. A telescope may be said to have a 'field of 3 degrees', meaning that parallel bundles of rays may pass through the telescope at angles up to 1.5 degrees from the optic axis. Sometimes the field semi-diameter, or the 'field semi-angle' is quoted. In this book, the field angle or *obliquity* of a parallel ray bundle will be denoted by θ.

3.6.5 The iris and the pupils of a system

These are defined as follows:

- **The iris**   There must be one stop which limits and defines the diameter of all the ray bundles that pass through the system.
- **The entry pupil** is the image, seen from the left (the side of the incoming light), of the iris of the system.[3] Optical elements of the system may form a real image of the iris inside

---

[3] Probably virtual, but real if the iris is in front of the first element, when the iris itself is the pupil.

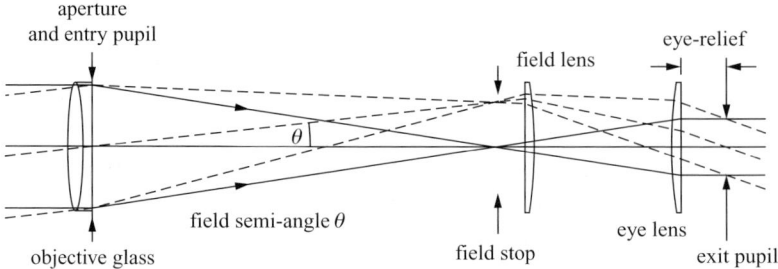

Figure 3.9 A refracting telescope. This is a complete optical instrument illustrating the concepts in this section. At the prime focus is a field stop to reduce scattered light and to define the edge of the field. The field lens follows and its purpose is to image the objective lens on to the exit pupil where the pupil of the eye will be located. The third lens is the eye lens and it recollimates the light emerging from the focal plane so that the image appears to the observer's eye to be at $-\infty$.

the system and at this image point there may be another iris, conjugate to the first. Either or both of these diaphragms may be regarded as the iris.
- **The exit pupil** is similarly defined as the image, real or virtual, viewed from the right of the iris.
- **The vignette** This occurs when there is no single defining stop controlling all the ray bundles that are to be transmitted. Figure 3.6 shows an example of this, where the two lens rims each control part of the bundle. The term was originally applied in the Victorian age to tasteful portrait photographs in which there was no sharp edge, but a general fading of the scene near the rim. It is highly undesirable in optical instruments and to prevent the resulting uneven illumination in the image, a stop is placed between the elements so that all ray bundles have the same size and shape. Nearly all optical instruments have such a defining stop.[4]

A refracting telescope (or a binocular), has the objective lens itself as both entry pupil and iris and the exit pupil is the image, usually about 2 cm to the right of the eye lens (the last lens of the system), of the objective. The diameter of this image is the diameter of the objective divided by the magnification of the telescope. It is at this point that the pupil of the eye must be placed if the whole field of view is to be seen.

This is an example of an important rule:

If two optical systems are to be coupled together the exit pupil of the first must be the entry pupil of the second. If both pupils are virtual, a relay lens must be inserted to image each on to the other.

---

[4] The Michelson Fourier-transform spectrometer is an exception.

3.7 Ray bundles

These are collections of rays which diverge from a point on the object, pass through different parts of the optical system and converge to form a point on the image. Their common property is that every ray of the bundle should have the same optical path length: if not, they interfere with each other. In practice they are not the same and a good optical image is formed if the differences are less than a couple of wavelengths or so. If you look with a microscope at an image in monochromatic light you can sometimes see the interference fringes.

3.7.1 Chief-rays

Every ray bundle has a central ray, the 'chief-ray', defined as follows:
The chief-ray of any bundle crosses the optic axis at the iris of the system. It follows that it also crosses the optic axis at the pupils provided they are *real* images of the iris, as for example in a telescope eyepiece.

The chief-ray serves to define the direction in which the bundle is propagating. If, for example, an instrument is to form a monochromatic image of a source using an interference filter, all the chief-rays must be parallel to the optic axis where they pass through the filter, since the wavelength transmitted by an interference filter depends on the angle of incidence of the light on it. This can be done in a telescope, for example, by placing at the focus of the objective a second lens, of the same focal length as the objective and the same diameter as the field of view desired. Since the objective lens of the telescope is the entry pupil, the chief-rays of all the ray bundles pass through the centre of the objective lens and are then collimated (made parallel) by this second lens placed at (or *near*, in practice) its focus. The interference filter follows, and then the eyepiece, camera or whatever is supposed to follow.

There are two precepts to bear in mind when designing an optical system:

- There is no such thing as a point source of light.
- There is no such thing as a parallel beam of light.

Both of these ideas are useful fictions when discussing Gaussian and geometrical optics, but in the study of photometry we must part company with them.

3.8 The Helmholtz–Lagrange invariants

It can be demonstrated graphically (Fig. 3.10) that the linear size of an image multiplied by the tangent of the angle subtended by the preceding and following

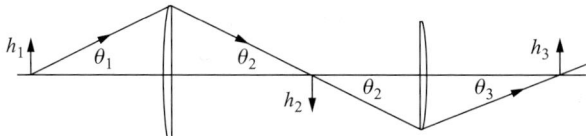

Figure 3.10 The Helmholtz–Lagrange invariant. It is a matter of elementary geometry that the product $h_i\theta_i$ is conserved throughout the system and there is a similar product in the plane through the optic axis perpendicular to this diagram. In practice the two invariants combine to give the law of conservation of étendue.

optical elements is a constant through a given centred system. To be precise,

$$n_1 h_1 \tan\theta_1 = n_2 h_2 \tan\theta_2 = n_3 h_3 \tan\theta_3 = \cdots = L,$$

where n_1 is the refractive index of the medium in the appropriate section. Usually $n = 1$.

If the field is anamorphic – as in a wide-screen cine-projector for example – there are two such invariants, perpendicular to each other, and each including the optic axis. The concept can be extended as follows:

The area of the field, multiplied by the solid angle subtended there by the pupil following it and by the square of the intervening refractive index, is a constant of the system.

It can alternatively be defined as the area of the pupil multiplied by the solid angle subtended by the preceding or following field-stop.

This constant is called the *étendue* of the system. It is a particularly valuable quantity to know when designing a photometer, since it immediately allows one to calculate the flux of light that will be accepted.

In practice it is usually sufficient to define it as

$$E = \frac{(\text{area of pupil}) \times (\text{area of field})}{(\text{distance from pupil to field stop})^2},$$

and it is generally quoted as cm^2 sterad.

Etendue is of particular importance when applied to spectrometers. It is also known as *luminosité* and generally denoted by the letter L. In a spectroscope for example it is easy to see that, when looking at a line emission spectrum, if the slit width is doubled the amount of light entering the eye is doubled. At the same time the resolving power – the ability to see two lines as separate – is halved. We shall see later that if resolving power is defined as $\lambda/d\lambda$ where $d\lambda$ is proportional to the width of the slit, then the product, LR, of étendue and resolving power is a constant for a particular design and $E = LR$ is defined as the efficiency of the instrument.

3.9 Surface brightness

The total power radiated by a source depends on its area and its *surface brightness*. This is defined in two ways. It is either:

(1) the power emitted per unit area per steradian normal to the surface; or
(2) the photon flux emitted per unit area per steradian normal to the surface.

It can be measured in watts per unit area per steradian, or photons per second per unit area per steradian. If the source is a *Lambert radiator* (it usually is not) then the total power radiated into 2π steradians is $\pi I S$, where I is the power per unit area per unit solid angle normal to the surface, and S is the area of the surface.

This follows from Lambert's law.

3.9.1 Lambert's law

This states that the power per unit solid angle emitted by a luminous surface is proportional to the cosine of the angle which the radiation makes with the surface normal.

Seen from a distance, a tilted surface of finite area S appears foreshortened to an area equal to $S \cos \theta$, and by Lambert's law, the radiation is similarly curtailed in intensity.

Thus the surface brightness of a perfect Lambert radiator is independent of the direction from which it is viewed. To a distant observer, for example, a radiating spherical surface will be indistinguishable from a uniformly illuminated plane disc.

The idea is convenient geometrically, but in nature and in the laboratory Lambert radiators are rare. The Sun, for example, is not a uniformly bright disc: its apparent surface brightness at the edge is less than half that at the centre, and moreover is wavelength dependent.

A perfect optical system will convert rays from a point source into a parallel beam. Since the power radiated by a source is proportional to its area, there will thus be no power in a parallel beam. The power will depend on the *divergence* of the beam, which in turn is measured by the angular diameter of the source at the entry pupil of the system.

3.10 Black body radiation

The most important practical consequence of Lambert's law is that if a Lambert surface emits a total of I photons per second per unit area, then the apparent surface brightness is I/π photons per second per unit area per steradian. A black body for example, emits σT^4 watts per unit area into 2π steradians. σ is Stefan's constant and T the absolute temperature. The surface brightness normal to the surface is then $\sigma T^4/\pi$ watts per unit area per sterad.[5]

3.10.1 Spectral power density

The total power emitted will depend on the nature of the emitting source. A black body radiates power as a function of wavelength according to the well-known Planck formula:

$$I(\lambda) = \frac{2\pi hc^2}{\lambda^5}\left(\frac{1}{e^{hc/\lambda kT}-1}\right), \qquad (3.2)$$

and this can be converted to a formula for the photon flux:

$$N(\lambda) = \frac{2\pi c}{\lambda^4}\left(\frac{1}{e^{hc/\lambda kT}-1}\right). \qquad (3.3)$$

The units in each case are watts (or photons per second) per unit area per unit wavelength range radiated into 2π steradians. N or I must be divided by π to give the spectral surface brightness.

Most sources do not radiate as black bodies. A polished reflecting surface has a low *emissivity* and neither emits nor absorbs radiation efficiently. Its emissivity may be a function of wavelength, $e(\lambda)$, which must multiply the Planck formula above.

The important property of surface brightness is that when multiplied by the étendue of the photometric instrument observing it, the result is the power or photon flux entering the detector.

Remember that the units of étendue are (area) × (solid angle) and that, multiplied by the units of surface brightness (flux per unit area per unit solid angle), yields flux.

These are the two factors involved in computing the yield of an optical photometer or spectrophotometer with a given source.

Worked example Consider a photometer with a detector area of 1.5 cm^2 at the focus of a lens 3 cm diameter and focal length 10 cm. Its étendue is $\pi(3/2)^2 \times 1.5/100 = 0.106$ cm^2 sterad. If it observes a distant extended source of luminosity 10^{10} photons cm^{-2} s^{-1} sterad^{-1} it will receive radiation at the rate of $10^{10} \times 0.106 = 1.06 \times 10^9$ photons s^{-1}.

Notice that the answer does not depend on the distance of the photometer from the source. In other words, if you sit facing a blank, uniformly illuminated wall you cannot tell how far you are from it: its surface brightness does not change as

---

[5] Not $\sigma T^4/2\pi$, because of the cosine factor.

you move away. Each cell of the retina is a separate detector with an étendue in the region of 10^{-6} cm^2 sterad. At a greater distance it will receive less light from unit area of the wall, but there will be a correspondingly greater area of the wall imaged on to each cell. The total light received by each cell is then unchanged.

However, when light arrives at a telescope from an apparently point object such as a star it is the aperture of the telescope which determines the signal at the detector, provided all the light from the object arrives at the same detector. The noise generated by the detector depends on the sensitive area[6] and it is usually desirable to have as small a detector area as possible. The numerical aperture or focal ratio of the telescope is then important, as the diameter of the diffraction-limited image is $1.22F\lambda$ where F is the focal ratio. Even when the size of the image is determined by 'seeing' to an angular size of about $1''$ of arc, the focal ratio still determines the diameter of the image at the detector. This is the point at which an aplanatic lens becomes important as a way of increasing the numerical aperture while retaining the spatial resolution.

---

[6] It is generally proportional to the *square root* of the sensitive area in a photoconductor, a bolometer or a photon counter such as a photomultiplier tube and is very temperature dependent.

4
Optical aberrations

When parallel rays pass through a centred, image-forming optical system infinitesimally far from the axis, they arrive at the corresponding *paraxial* image point. Its position is computed using the Gaussian optical equations.

It is an unfortunate fact that in general parallel rays coming through the system at the outer parts of the apertures or pupils do not arrive at the same image point as the paraxial rays. They arrive in the vicinity of the paraxial image point and their separations from it define the *aberrations* of the system. Rays which pass the edges of the pupils are called the *marginal rays* and the separation of these rays from the paraxial image point can be calculated using aberration theory.

In instrument design we arrange the various components of the system so that the aberrations are reduced to zero where possible and minimised where not. The whole theory is non-linear; the various causes of aberration interact with each other and the resulting complication is often deplorable. To ease the analysis we simplify things by treating each aberration as though the others were all absent.

Optical designers are sometimes eccentric and the routes to their designs are often highly individual and hard to follow. Conrady, for example, is regarded as the father of modern optical design but his notation is not to be recommended. In any case, for plain spectrograph optics we need only a small subset of the theory and this is laid out below in reasonably familiar language.

There are two principal ways of reducing the aberrations of an instrument.

The first is by altering the shape of a lens,[1] by thickening and *bending* it by changing its radii of curvature, c_1 and c_2, without altering its power $(n-1)\Delta c$.

The second is by placing *stops* so as to fix the position of the pupils[2] of the system, since the aberration produced by an optical surface depends on the distance of its

---

[1] The shape of a mirror can be changed by 'figuring' it, thereby changing its shape from spherical to paraboloidal, ellipsoidal or to a fourth-order curve.
[2] Strictly the *iris* although the word 'pupil' is used commonly and carelessly in this connection.

vertex from its pupils. The position of the iris diaphragm in a camera lens, for example, is carefully chosen as part of the design.

There are various theorems which guide the instrument designer and there are ray-tracing computer programs which verify (or not!) the calculations which have been made. The whole subject is an art rather than a science and in the fine detail, experience tends to be the guiding principle as much as cold analysis.

4.1 The Seidel aberrations

Gaussian optics – the first step in instrument design – relies on the approximations $\sin\theta \simeq \theta$; $\cos\theta \simeq 1$. From this the approximate positions of image planes, magnifications and so on can be calculated. This is the 'zero-order' theory.

Seidel theory, sometimes called 'first-order theory' or 'third-order theory', makes the approximations $\sin\theta \simeq \theta - \theta^3/6$, $\cos\theta \simeq 1 - \theta^2/2$ and $\tan\theta \simeq \theta + \theta^3/3$. Analysis of ray paths then shows the marginal rays of a bundle passing at calculable distances from the paraxial image point.

NB Seidel theory is not exact, but it brings the design close enough to a starting point for ray tracing, when experimental refinement of the design becomes possible, formerly by hand calculation using seven-figure logs and trigonometric tables, but these days by computer-iterative ray tracing.

The aberrations can be distinguished by their severity and by the way in which they depend for their amplitude on the *semi-aperture y* and *field* θ of the system. The derivation of the equations describing them is in many textbooks and they are distinguished by various coefficients, of which we shall need four: **B, F, C** and **D**. In diminishing order of importance the aberrations are detailed below.

4.2 Zero-order aberration

Defect of focus The rays of a bundle arriving at the Gaussian image point all pass through the same point, but this point is before or after the image plane. This aberration is easily corrected by moving the image plane to its proper position.

4.3 First-order aberrations

4.3.1 Spherical aberration

This is caused by employing optical elements with spherical surfaces, and affects all image points in the same way. If the Gaussian image point is on the optic axis, the marginal rays cross the axis before or after the paraxial rays. The distance along the axis from the paraxial image point to the marginal image point is called the

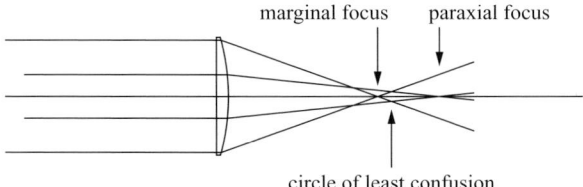

Figure 4.1 Spherical aberration. The longitudinal aberration is the distance between the points where paraxial rays and marginal rays of a collimated bundle cross the optic axis. Of more practical importance is the diameter of the circle of least confusion, which is the quantity which defines the spatial resolution of the lens.

longitudinal spherical aberration and in the Seidel approximation it is proportional to the focal length and varies as the square of the aperture. For a lens or mirror of focal length f the usual formula for longitudinal spherical aberration is

$$\Delta_{\text{sph}} = \mathbf{B} y^2 f, \tag{4.1}$$

where y is the *semi-aperture* (the radius of the lens if it is a simple lens) and f the focal length. **B** is the *spherical aberration coefficient* and its value depends on the *bending* of the lens. There are formulae in Appendix 1 for various circumstances. **B** is well known to be zero for a paraboloidal reflector.

It is generally more useful to know the diameter of the *circle of least confusion*; that is, the diameter of smallest image patch that can be achieved when the image plane is appropriately adjusted. The formula for this is

$$\Delta_{\text{colc}} = \mathbf{B} \frac{y^3}{2}. \tag{4.2}$$

The patch is also known as the *blur circle*.

Spherical aberration can be corrected in a simple lens by making it an *achromatic doublet*. This lens has two components in contact, generally cemented together, of different types of glass, one component having positive and the other negative power. They are chosen so that the focal length of the combination is the same for two different wavelengths, that is for two different refractive indices. By 'bending' the combination, the spherical aberration of the two components can be made equal and opposite, and as the components are in contact the two aberrations add simply. This process will be discussed in detail later.

4.3.2 Coma

This is regarded as the great Satan of optical design, and it is a rule of thumb that, if coma can be eliminated, the lesser aberrations tend to take care of themselves, or at least are less troublesome to reduce to acceptable levels. Coma depends on the

4.3 First-order aberrations

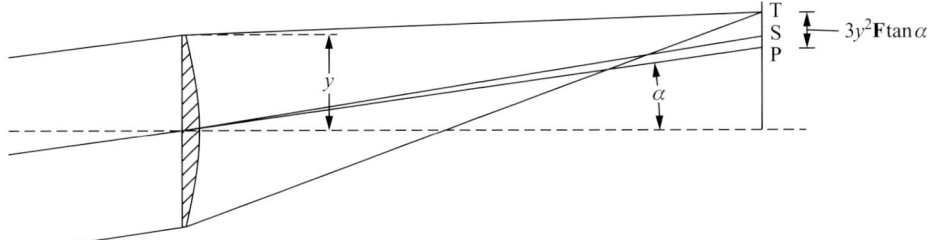

Figure 4.2 The meridional or tangential coma of a lens. In the absence of other aberrations, the marginal rays of an oblique bundle in the plane of the optic axis meet at the tangential image point T, which generally is not on the oblique axis – the line joining the Gaussian image point P to the lens vertex. Tangential coma is then measured as the distance in the focal plane from point P to the tangential focus point T. Sometimes it is measured as the angle subtended by this distance at the lens vertex. The oblique marginal rays perpendicular to this diagram meet at S, the sagittal focus, and in the Seidel approximation the distance PS is always 1/3 of the distance PT.

field angle of the chief-ray of a bundle and we therefore consider *oblique* bundles of rays where the chief-ray is inclined at an angle θ to the optic axis.

In an oblique ray bundle, the paraxial rays arrive as before at the paraxial focus. If there is no spherical aberration, the marginal rays converge to the same focal surface, but they arrive at different points on that surface. Two pairs of marginal rays are particularly important and may be said to define the amount of coma.

Tangential coma

When the two marginal rays in the plane of the optic axis meet at a point which is not on the paraxial ray, the perpendicular distance of this point from the paraxial ray is the tangential coma. The height of the coma patch is

$$\Delta_{\text{coma}} = 3y^2 \mathbf{F} \tan \theta. \tag{4.3}$$

F, the coma coefficient, and **B**, the spherical aberration coefficient, are the two most important design parameters in a spectrograph. Appendix 1 gives formulae for calculating **B** and **F** for single lenses and mirrors.

Sagittal coma

In a simple case, the two marginal rays from the ends of the diameter perpendicular to the plane of the chief-ray arrive at a point between the paraxial focus and the point where the meridional marginal rays meet. Its distance from the paraxial focus is one-third of the distance to the meridional marginal focus and, by symmetry, is on the line joining them. Other pairs of marginal rays arrive at points on a circle, of which the meridional and sagittal foci are at the ends of a diameter. Ray pairs from each annulus of the lens form circles of different diameter and the resulting

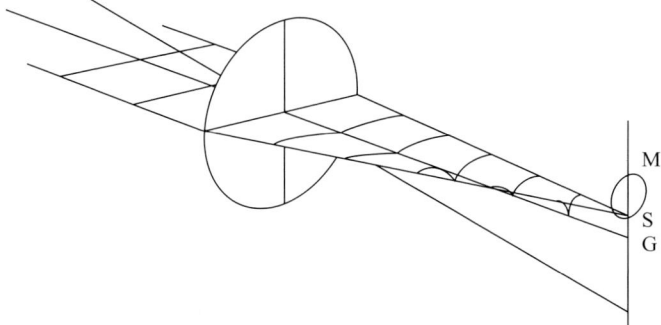

Figure 4.3 The sagittal coma of a lens. This is defined by the point S, where marginal rays in the plane normal to the plane of Fig. 4.2 meet. Pairs of marginal rays from other mutually antipodal points on the rim of the lens meet on a small circle which has the tangential and sagittal foci at the ends of a diameter. Ray pairs from other, smaller annuli form similar, smaller circles all lying cradled between two lines of 60° inclination radiating from the Gaussian focus G.

coma patch is triangular with its apex at the paraxial focus and an apex angle of 60°. The illumination is uneven, being fainter nearer the outer edge, and it is from this comet-like form that the aberration gets its name.

4.3.3 Astigmatism

This aberration is closely linked to field curvature and the two are usually considered together. Like the **B** and **F** coefficients of spherical aberration and coma, the other two are **C** which vanishes for zero astigmatism and **D** which vanishes if the field is flat.

Suppose that a distant wheel-like object in the form of concentric black circles and radial lines on a white field, centred on the optic axis and perpendicular to it, is imaged by the system. If astigmatism is present, there are two focal surfaces, both spherical. One of them will show the rings in focus and the other the radial lines. The two surfaces have a common tangent at the Gaussian focus.

In neither case does a point on the object correspond to a point in the image.

An annular parallel bundle of rays inclined to the optic axis will first converge to form a short line image; a line which, if projected, will cut the optic axis: and the bundle will subsequently converge to a short line image perpendicular to the former. This is positive astigmatism. The position of the first short line is the *sagittal* focal surface – the spokes of the wheel – and the other the *meridional* or *tangential* surface – the rim of the wheel. The two do not necessarily happen in the order given here. Astigmatism, like coma, may be negative.

In classical prism and grating spectroscopes it is the tangential surface which chiefly concerns us. This is where the spectrum lines are focused.

4.3 First-order aberrations

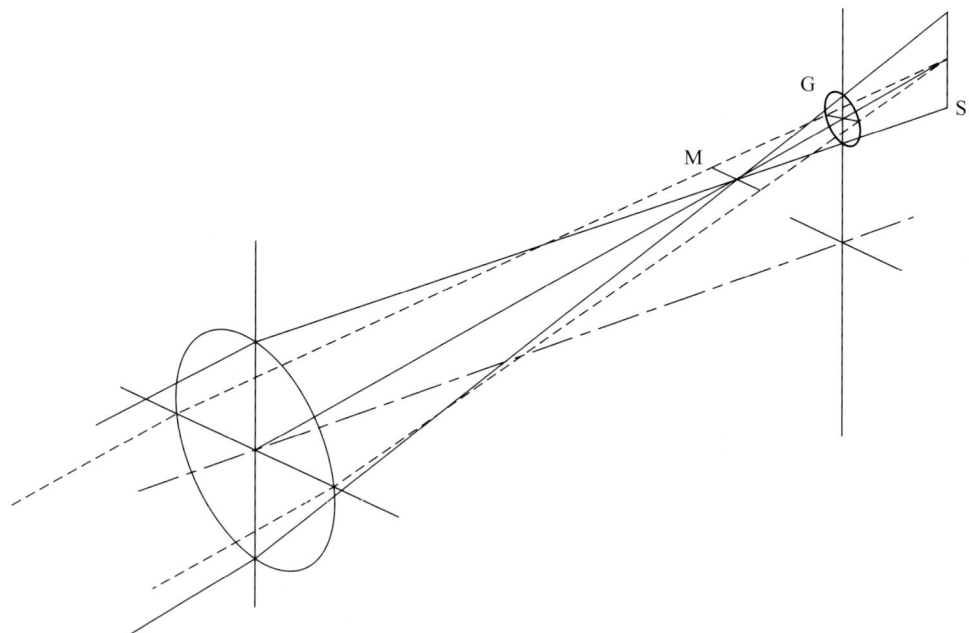

Figure 4.4 Astigmatism. In the absence of coma the tangential and sagittal ray fans meet at different points along the oblique axis forming line images of a point object. Where the rays of the tangential bundle meet, the tangential line image M is perpendicular to the optic axis, and at the sagittal focus the sagittal line image S if produced, would cut the optic axis. There is a 'circle of least confusion' G, lying between the two foci. In spectroscopy it is the tangential focus which is almost always required. Consequently a spectrum line is not a precise image of the entry slit but the aberration, being along the slit direction, does not interfere with the spectral resolution.

Astigmatism, like coma, is an aberration of the field and vanishes on the optic axis. In order for it to exist at all there must simultaneously be *field curvature*.[3] One of the two field curvatures may be zero. In a grating spectrograph, for example, the tangential field may be flat and the entry slit will then be imaged, in various colours, on to a plane. The image of a point on the slit is a short line lying along the direction of the spectrum line image. Thus we do not notice it unless the slit is very long or the spectral resolution very high.

4.3.4 Astigmatism and field curvature

Field curvature on its own means that the image of a scene at infinity will not lie on a plane but on a sphere which has its centre on the optic axis and which cuts the

---

[3] Although the latter may exist even in the absence of astigmatism.

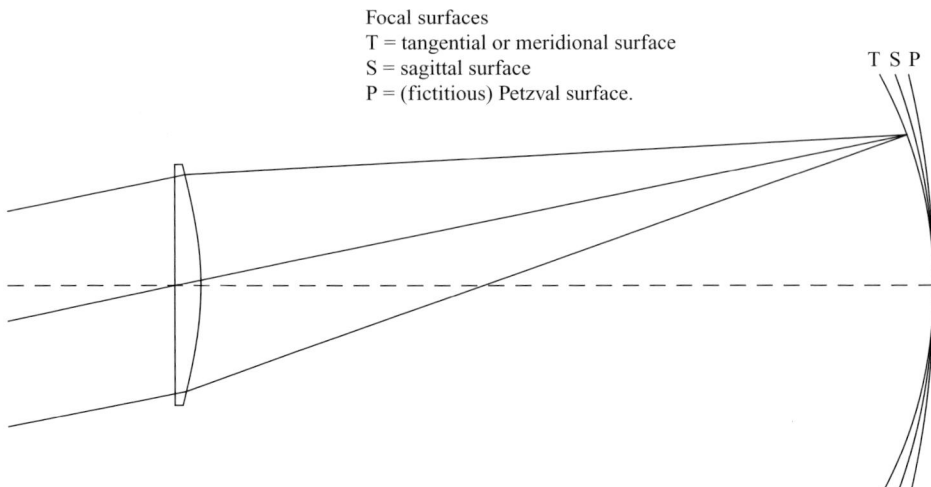

Figure 4.5 Field curvature. This is inseparable from astigmatism as the tangential and sagittal images lie on two different spherical surfaces. If there is no astigmatism the two surfaces are the same, and if field curvature is also eliminated the field is flat. There is a third, fictitious surface called the Petzval surface, the radius of which is defined in the text, and the order T-S-P of the curvatures is always preserved.

optic axis at the Gaussian focus. The field curvature is defined by the radius of this sphere or by its reciprocal.

This sphere may be convex or concave towards the lens, i.e. its radius of curvature may be positive or negative. For many purposes, such as photography, CCD imaging, etc., it is highly desirable that the radius be infinite so that the focal surface is a plane. There is one notable exception: the Schmidt camera, where all aberrations except field curvature are perfectly corrected and a plane glass photographic plate is bent to form a spherical surface (of fairly large radius, although glass is more elastic than most people realise). The symbol c is used in some books for field curvature as well as surface curvature. To avoid confusion lower-case c is reserved for surface curvatures of lenses and mirrors, and capital $C_{(\text{suffix})}$ is used for field curvatures. C_P, C_s and C_t refer to Petzval, sagittal and tangential field curvature respectively. **C** refers to the astigmatism coefficient and, like the other aberration coefficients, is always in bold font.

4.3.5 Distortion

This is characterised by a fifth coefficient **E**, and occurs when the position of the principal plane depends on the field angle, so that the focal length is a function of the field angle. In the absence of all other aberrations the focal length may either increase or decrease with field angle. If it increases with field angle the result is

4.4 Theorems

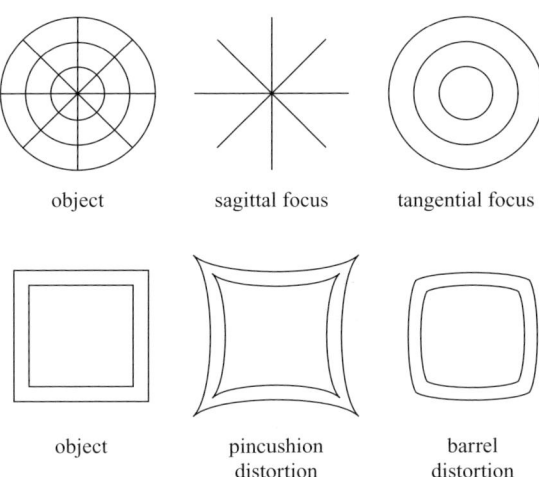

Figure 4.6 Images of a wheel-object, showing the practical effects of astigmatism and distortion in the images which they produce. Distortion, usually barrel distortion, is inevitable in lenses corrected to give very wide fields but is small or absent in so-called 'symmetrical' lenses such as the double Gauss camera objective.

'pincushion' distortion and if it decreases it produces 'barrel' distortion. In other words the image of a square object is shaped as the name implies. Barrel distortion is difficult to avoid with wide-angle lenses.[4]

4.4 Theorems

4.4.1 Fraunhofer's theorem

This is an important theorem which shows how it is possible to place a stop to control the coma of a system. If the entry pupil of a simple lens is at the lens itself, the pupil distance is zero and the coma coefficient is written as $\mathbf{F}_0$.

Fraunhofer's theorem is that if the entry pupil is at a distance Z from the principal point, the coma coefficient becomes

$$\mathbf{F} = (\mathbf{F}_0 + Z\mathbf{B}). \tag{4.4}$$

It follows then that if a stop is placed a distance Z from the vertex, given by $Z = -\mathbf{F}_0/\mathbf{B}$, the lens will be free from coma. The distance is negative and so, by the Cartesian sign convention, the stop must be to the left of the vertex.

Consider for example a spherical interface between glass and air, with a stop at the centre of curvature of the glass sphere.[5] Parallel bundles of rays, no matter

---

[4] Take especial care when choosing a lens for a Fabry–Perot spectrograph, where the dispersion is radial and only a camera lens with freedom from distortion gives a true square-law wavelength scale.
[5] This example is repeated in Appendix 1 with attendant mathematical details.

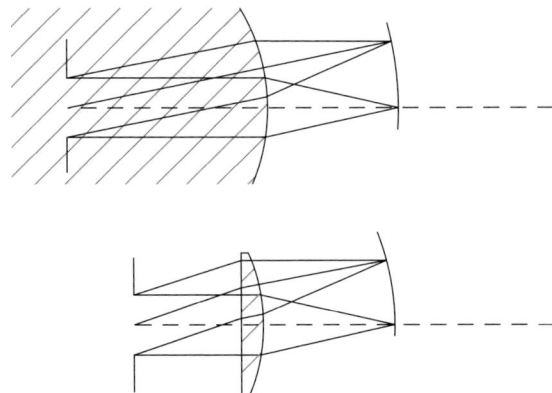

Figure 4.7 Controlling coma with a stop. In the upper figure a stop is placed at the centre of curvature of a glass sphere so that collimated ray bundles coming from the left are always symmetrical and are imaged with spherical aberration but no coma on a spherical focal surface. In the lower figure the sphere has been truncated to give a plano-convex lens with a plane first surface, and the stop position is moved so that the ray bundles reaching the second surface are unchanged. A stop at this position in front of a plano-convex lens will thus eliminate coma.

in which direction they are travelling, are symmetrical about their chief-ray since, because of the spherical symmetry, there is no unique optic axis. There must then be no coma when the bundle comes to a focus after refraction, although there will be spherical aberration.

Now place a second, plane, surface to the left of the spherical surface to make a plano-convex lens. The stop, now in air to the left of the lens, can be moved a distance $R[(n-1)/n]$ towards the lens so that parallel bundles of rays from it will arrive at the second spherical surface just as if they had come from its centre of curvature. To put it another way, the stop should be a distance $f[(n-1)/n]$ from the plane surface of the lens. Again there will be no coma and the stop is at the position required by Fraunhofer's theorem.

This is an elementary example showing the underlying physics: the theorem is of more general use than this.

A word of warning: coma can only be corrected with a stop if there is some spherical aberration. It happens occasionally that it is worth leaving some spherical aberration in the system deliberately in order that coma may be removed. For example, the 'Ross corrector', which was used in large astronomical telescopes with paraboloidal (and therefore spherical-aberration-free) mirrors, did just this. The pupil distance is so large (from the telescope primary to the vicinity of the focal plane) that the introduction of a totally insignificant amount of spherical aberration is sufficient to allow coma correction over a large area on the image.

On its own a singlet lens has no coma provided that its surface curvatures are connected by

$$\frac{c_2}{c_1} = \frac{(n^2 - n - 1)}{n^2},$$

and typically, for $n = 1.516\,728$, $c_2 = -0.94 c_1$ will give coma below the diffraction limit.

If the refractive index is $1.618\,034$ – the 'golden section' of the pyramid-builders and Greek architects – the coma of a plano-convex lens is zero when the first surface is convex. This is no gift of the gods however: all the other aberrations are there in full measure!

4.4.2 Petzval's theorem

When astigmatism is present the tangential focal surface is a sphere of curvature $C_t = 4\mathbf{C} + 2\mathbf{D}$, and the sagittal surface curvature is $C_s = 2\mathbf{D}$.

Petzval's theorem relates the two coefficients $\mathbf{C}$ and $\mathbf{D}$ by the equation

$$\mathbf{C} - \mathbf{D} = \sum_i \frac{1}{n_i f_i}, \tag{4.5}$$

where n_i is the refractive index of the ith component. This called the 'Petzval sum' and denoted by the symbol $\sum P_z$.

A surface can be defined, the 'Petzval surface', of which the curvature, C_P, is $-2(\mathbf{C} - \mathbf{D})$ and the two astigmatic surfaces have curvatures given by

$$C_t = C_P + \frac{3}{2}\delta C,$$
$$C_s = C_P + \frac{1}{2}\delta C.$$

$\mathbf{C} - \mathbf{D} = 0$ is the condition for freedom from astigmatism, and $\mathbf{C} = \mathbf{D} = 0$ are the conditions for freedom from both astigmatism and field curvature.

4.5 Aberration coefficients for mirrors

These are collected here as they can be given in closed form and are most likely to be used in spectrograph design. Formulae from which the equivalent coefficients for lenses may be computed are collected in Appendix 1. When using them bear in mind that these are only first-order improvements on Gaussian optics and are not exact. They give reasonable results at apertures slower than $\sim F/12$ and field angles up to $\sim 5°$ and are adequate starting points for iterative ray-tracing programs.

e is the excentricity of the conic of which the mirror is the surface of revolution. **Z** is the vertex–pupil distance and is negative when it is to the left of the vertex.

The spherical aberration coefficient is

$$\mathbf{B} = (1 - e^2)/8f^2. \tag{4.6}$$

The coma coefficient is

$$\mathbf{F} = \frac{1}{4f} + \frac{Z(1-e^2)}{8f^2}. \tag{4.7}$$

The astigmatism coefficients are

$$\mathbf{C} = \frac{1}{2f} + \frac{Z}{2f^2} + \frac{Z^2(1-e^2)}{8f^3}, \tag{4.8}$$

$$\mathbf{D} = \frac{Z}{2f^2} + \frac{Z^2(1-e^2)}{8f^3}. \tag{4.9}$$

The tangential surface curvature is

$$C_t = 4\mathbf{C} + 2\mathbf{D} = \frac{2}{f} + \frac{3Z}{f^2} + \frac{3}{4}\frac{Z^2(1-e^2)}{f^3}. \tag{4.10}$$

The sagittal surface curvature is

$$C_s = 2\mathbf{D}. \tag{4.11}$$

From this it is a short step to see that a spherical mirror is free from astigmatism if $Z = -2f$. The field curvature is then $-1/f$. These are the conditions for a Schmidt camera.

4.5.1 The focal surfaces of spectrographs

These last two expressions are particularly important for Čzerny–Turner spectrographs. The tangential field is flat when

$$\frac{2}{f} + \frac{3Z}{f^2} + \frac{3}{4}\frac{Z^2(1-e^2)}{f^3} = 0. \tag{4.12}$$

When spherical mirrors are used the grating–mirror distance, $Z, = -0.843f$. (The alternative value, $Z = -3.25f$, can safely be ignored.) If the mirrors are paraboloids, then $Z = -0.666f$.

4.5.2 Chromatic aberration

A simple single-element lens has a focal length which depends on its refractive index, *n*, and refractive index generally increases with decreasing wavelength. In *crown glass* this variation is comparatively low and in *flint glass* it is higher. There is no discontinuous change in this rate of variation with glass type and a quantity called the *partial dispersion* or *Abbé v-value* is used to measure it. The partial dispersion of a glass is defined by

$$v = \frac{n_D - 1}{n_F - n_C},$$

where n_C, n_D and n_F are the refractive indices at 656.3 nm, 589.3 nm and 486.1 nm respectively.[6] Typical values are

Glass	Index	v-value
Crown (BK7)	1.516 728	64.2
Flint (SF12)	1.648 143	33.8

and there is an arbitrary division between crown and flint glasses at $v = 50$. Notice that the flints, with their higher dispersions, have the lower v-values.

The consequences of this change of focal length with colour are:

- The image of a white luminous point object is a coloured point surrounded by a disc of the complementary colour. This is known as *longitudinal colour*. The actual colours depend on where the focusing screen is placed.
- The red image of a finite object is larger than the blue image (*lateral colour*).

Correction of chromatic aberration then requires both that the focal length be the same for different wavelengths and that the principal planes shall coincide for different wavelengths. Only then will the images of an object be superimposed on the focusing screen.

4.6 The achromatic doublet

Two elements, of two glasses with different refractive indices and dispersions, can be combined to make a doublet lens which has the same focal length for two different wavelengths. The parameters available are the two 'bendings' (that is the actual shapes of the two elements), the two refractive indices and the two dispersions; although these latter, being physical properties, depend on the glass types available.

The algebra is elementary. From the Gaussian formula, $P = (n - 1)C$, we find

$$P = P_1 + P_2 = (n_1 - 1)C_1 + (n_2 - 1)C_2$$
$$= (n'_1 - 1)C_1 + (n'_2 - 1)C_2,$$

---

[6] These wavelengths correspond to the similarly lettered Fraunhofer lines in the solar spectrum.

where n'_1 and n'_2 are the refractive indices at the second wavelength. Then

$$\Delta n_1 C_1 = -\Delta n_2 C_2,$$
$$P = (n_1 - 1)C_1 - (\Delta n_1/\Delta n_2)(n_2 - 1)C_1$$
$$= \Delta n_1 C_1 \left(\frac{n_1 - 1}{\Delta n_1} - \frac{n_2 - 1}{\Delta n_2} \right) \simeq C_1 \Delta n_1 (v_1 - v_2)$$

and the $\simeq$ sign indicates that an intermediate wavelength has been used for the factors $(n_i - 1)$.

After a few more lines of algebra you will find

$$P_1 = \left(\frac{v_1}{v_1 - v_2} \right) P$$

and

$$P_2 = -\left(\frac{v_2}{v_1 - v_2} \right) P.$$

From these two equations the curvatures of the two elements can be computed. However, nothing has been said about the individual curvatures of the four surfaces. These can be adjusted to make the spherical aberrations equal and opposite.[7] A minor caution, however: the chromatic aberration is not perfectly corrected. Although the formulae above provide the same focal length for the two colours, the principal planes and hence the focal planes are not *exactly* coincident and small longitudinal colour remains. At F/10, though, this is comparable with the diffraction disc and is not significant. Yet the ray tracer will correct for this also and will remove one or other of these two aberrations. Finding tool radii to match is another problem entirely and best left to the lensmakers.

The search for a respectable flat field of 10°–20° has been going on since the invention of photography and has resulted in the elaborate photographic objectives discussed by Kingslake,[8] to be found especially on 35 mm cameras of quality.

---

[7] In the old days this would have been done by hand using aberration formulae. Today the iterative ray-tracing program does it instantly.
[8] R. Kingslake, *A History of the Photographic Lens* (San Diego: Academic Press, 1989).

5
Fourier transforms: a brief revision

5.1 Fourier transforms

Fourier transforms and Fourier methods generally are such a powerful tool in describing physical optics and spectrography that a familiarity with them is a sine qua non of the optical designer. The spectrograph is the physical manifestation of a Fourier transformer. It takes an incoming signal, a function of time, and displays the frequencies present in it. It performs in fact the *power* transform of the signal, a concept which will be described later.

This chapter is chiefly to introduce the reader to the notation and to summarise the more important properties and theorems of Fourier theory. The notation in particular differs from that of mathematicians because we prefer to deal with real, measured quantities rather than the angular variables which they usually employ. The conjugate variables x and p traditionally used in Fourier theory may by replaced by t and ν, that is time in seconds and frequency (*not* angular frequency) in hertz, and functions of these two variables are connected by the Fourier inversion theorem:

$$\phi(\nu) = \int_{-\infty}^{\infty} f(t) e^{-2\pi i \nu t} \, dt,$$

$$f(t) = \int_{-\infty}^{\infty} \phi(\nu) e^{2\pi i \nu t} \, d\nu,$$

which we usually write symbolically as

$$f(t) \rightleftharpoons \phi(\nu).$$

Note that this particular notation frees the definitions from the normalising constants which otherwise appear in front of the integral sign.

In geometrical optics we also have to deal with a spatial dimension and its inverse, spatial frequency. On a grating for example, we may have a 'ruled width' x measured in centimetres and the fineness of the grating measured in lines per

42   *Fourier transforms: a brief revision*

centimetre p. Similarly in photography, the resolution of a lens or a photographic emulsion may be defined in line-pairs per centimetre, and by convention this uses the symbol s.

5.2 Theorems

Various elementary theorems follow here and for brevity are given without proof. Proofs are generally simple and to be found in any treatise on Fourier theory.[1]

5.2.1 The addition theorem

$$f_1(t) + f_2(t) \rightleftharpoons \phi_1(\nu) + \phi_2(\nu). \tag{5.1}$$

5.2.2 The multiplication theorem

$$\int_{-\infty}^{\infty} \phi_1(\nu) \phi_2^*(\nu) \, d\nu = \int_{-\infty}^{\infty} f_1(t) f_2^*(t) \, dt, \tag{5.2}$$

where the asterisk denotes a complex conjugate. An important special case of this is.

5.2.3 Parseval's theorem

$$\int_{-\infty}^{\infty} |\phi(\nu)|^2 \, d\nu = \int_{-\infty}^{\infty} |f_1(t)|^2 \, dt.$$

5.2.4 The shift theorem

$$f(t + t_0) \rightleftharpoons \phi(\nu) e^{2\pi i \nu t_0}, \tag{5.3}$$

where t_0 is a constant. Its corollary is important:

$$f(t + t_0) + f(t - t_0) \rightleftharpoons 2\phi(\nu) \cos 2\pi \nu t_0.$$

5.3 Convolutions

These are conveniently described by a spectroscopic example. No spectrometer is 'perfect'. It it were and if monochromatic light really existed then the spectrum

---

[1] The exact and rigorous proof of the fundamental inversion theorem is anything but simple and to be found for example in Titchmarsh, a bible to be consulted only by the most didactic reader.

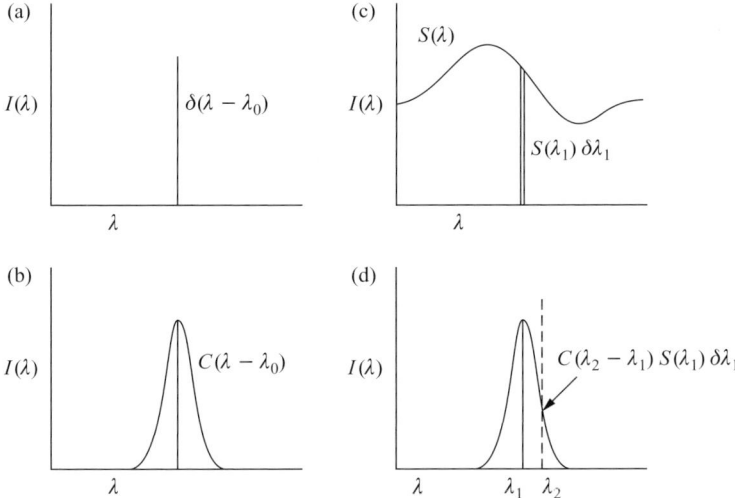

Figure 5.1 Convolution of an instrumental profile with a continuous spectrum. Each element of the spectrum is treated as a monochromatic emission line which is converted by the shortcomings of the spectrograph to the shape of the instrumental profile. The sum of all these conversions is added to give the observed spectrum.

of the light would be a Dirac delta function $A\delta(\lambda_0)$ at the appropriate place in the spectrum, where A is the intensity. In the real world the spectrometer produces a curve of intensity against wavelength which we denote by $AI(\lambda - \lambda_0)$ and which either peaks at λ_0 or has its centre of gravity at λ_0. The function $I(\lambda)$ is called the *instrumental profile* of the spectrometer and $\int_{-\infty}^{\infty} I(\lambda)\,d\lambda = 1$. The instrumental profile in general does not hold its shape over the whole spectrum but the shape changes slowly enough with wavelength that it serves to demonstrate the idea of a convolution.

Consider now a real spectrum, $S(\lambda)$, and in particular an infinitesimal part, of intensity $S(\lambda_1)\,d\lambda_1$ at wavelength λ_1. This approximates to a δ-function and will be converted by the defects of the spectrometer to a function $S(\lambda_1)\,d\lambda_1 I(\lambda - \lambda_1)$ and this in turn will contribute an infinitesimal amount of power $S(\lambda_1)I(\lambda_2 - \lambda_1)\,d\lambda_1$ apparently at a wavelength λ_2. The apparent spectrum, which is the total power appearing at λ_2 from all parts of the true spectrum, is then the integral

$$P(\lambda_2) = \int_0^{\infty} S(\lambda_1)I(\lambda_2 - \lambda_1)\,d\lambda_1,$$

and this is a practical example of the mathematical operation of convolution:

$$c(x') = \int_{-\infty}^{\infty} f_1(x)f_2(x' - x)\,dx, \qquad (5.4)$$

usually written symbolically as

$$c(x') = f_1(x) * f_2(x).$$

The operation can be visualised by drawing the function $f_1(x)$ on a sheet of paper, drawing $f_2(x)$ on transparent paper, turning the drawing over and laying it on top of the sheet with $f_1(x)$ and sliding it along, integrating the product of the two as you go. Each integral gives one point on the convolution.

5.3.1 The convolution theorem

This theorem is a most important and powerful one, particularly in Fraunhofer diffraction theory, and it greatly simplifies the mathematics of diffraction.

It states that if $c(x)$ is the convolution of $f_1(x)$ and $f_2(x)$ as above, then their respective Fourier transforms $\gamma(p)$, $\phi_1(p)$ and $\phi_2(p)$ are related by

$$\gamma(p) = \phi_1(p)\phi_2(p). \tag{5.5}$$

In words: the Fourier transform of a convolution is the product of the Fourier transforms of the components.

5.3.2 Convolution algebra

There is an algebra of convolutions which can be used to manipulate them. It is not difficult. The rules are as follows:
The commutative law

$$f_1(x) * f_2(x) = f_2(x) * f_1(x).$$

The associative law

$$f_1(x) * [f_2(x) * f_3(x)] = [f_1(x) * f_2(x)] * f_3(x).$$

The distributive law

$$f_1(x) * [f_2(x) + f_3(x)] = f_1(x) * f_2(x) + f_1(x) * f_3(x),$$

but note that convolutions and multiplications must be kept separate. The distributive law does not hold in a mixture of them:

$$f_1(x) * [f_2(x) f_3(x)] \neq [f_1(x) * f_2(x)] f_3(x).$$

Nevertheless the algebra can be extended. For example,

$$[f_1(x) f_2(x)] * [f_3(x) f_4(x)] \rightleftharpoons [\phi_1(p) * \phi_2(p)][\phi_3(p) * \phi_4(p)]$$

where Greek and Roman letters denote Fourier pairs as before.

5.4 The Wiener–Khinchine theorem

We first of all define spectral power density (SPD), which is what a spectrometer measures.

The autocorrelation function of a function $f(t)$ is similar to the convolution of a function with itself, and is defined as

$$A(\tau) = \lim_{T \to \infty} \frac{1}{2T} \int_{-T}^{T} f(t) f(t + \tau) \, dt,$$

where the factor before the integral implies a time average.

If the Fourier pair of $f(t)$ is $\phi(\nu)$, the shift theorem gives the Fourier transform of $f(t + \tau)$ as $\phi(\nu) e^{2\pi i \nu \tau}$ and the multiplication theorem then gives

$$\int_{-T}^{T} f(t) f(t + \tau) \, dt = \int_{-\infty}^{\infty} \phi^*(\nu) \phi(\nu) e^{2\pi i \nu \tau} \, d\nu = \int_{-\infty}^{\infty} |\phi(\nu)|^2 \, e^{2\pi i \nu \tau} \, d\nu.$$

On the (reasonable) assumption that the power is zero before the beginning and after the end of the signal receipt, we define the SPD to be $G(\nu)$ given by

$$G(\nu) = \frac{|\phi(\nu)|^2}{2T}.$$

The Wiener–Khinchine theorem states, as in the equation above, that the SPD is the Fourier transform of the autocorrelation function of the signal $f(t)$:

$$A(\tau) \rightleftharpoons G(\nu),$$

or, more practically,

$$\frac{1}{2T} \int_{-T}^{T} f(t) f(t + \tau) \, dt \rightleftharpoons \frac{|\phi(\nu)|^2}{2T}. \tag{5.6}$$

5.5 Useful functions

There are some mathematical expressions which recur frequently in physical science in conjunction with their Fourier transforms or Fourier *pairs*. Some of them are described here.

5.5.1 The 'top-hat' function

This is a function which is zero from $-\infty$ to $-a/2$ and from $a/2$ to ∞. Between $-a/2$ and $a/2$ it has the value 1. It is generally denoted by $\Pi_a(x)$.

Its Fourier transform is

$$\int_{-\infty}^{\infty} \Pi(x) e^{2\pi i p x}\, dx = \int_{-a/2}^{a/2} e^{2\pi i p x}\, dx = \frac{e^{\pi i p a} - e^{-\pi i p a}}{2\pi i p}$$

$$= a \frac{\sin \pi p a}{\pi p a} = a\,\mathrm{sinc}(\pi p a),$$

and the so-called 'sinc' function is much encountered in physics and engineering, not least because data strings have a finite length and the length in turn limits the frequency resolution. This is because the observed spectrum – the Fourier transform of the data – is the convolution of the true spectrum with the sinc function.

5.5.2 The Gaussian function

The Fourier transform of the Gaussian function e^{-x^2/a^2} is obtained directly by 'completing the square' of the exponent in the integrand:

$$\int_{-\infty}^{\infty} e^{-x^2/a^2} e^{2\pi i p x}\, dx = e^{-\pi^2 p^2 a^2} \int_{-\infty}^{\infty} e^{-(x/a + \pi i p a)^2}\, dx$$

$$= a\sqrt{\pi}\, e^{-\pi^2 p^2 a^2}.$$

The full width at half maximum (FWHM) of a Gaussian is $2a\sqrt{\ln 2}$, and numerically is $1.386a$ where a is the 'width parameter'. The convolution of two Gaussians of width parameters a and b is another Gaussian of width parameter $\sqrt{a^2 + b^2}$, i.e. the Pythagorean sum of the two width parameters.

5.5.3 The Lorentz function

When a damped electrical oscillator decays the amplitude it radiates varies with time according to $f(t) = A_0 e^{2\pi i \nu_0 t} e^{-t/\tau}$, where ν_0 is its natural oscillation frequency and τ the decay constant.

The spectrum of the oscillation is the Fourier transform of this:

$$\overline{A}(\nu) = A_0 \int_0^{\infty} e^{2\pi i \nu_0 t} e^{-t/\tau} e^{-2\pi i \nu t}\, dt,$$

and the integral lower limit is 0 because the oscillation is deemed to begin at this time. The integral is then

$$\overline{A}(\nu) = \frac{A_0}{2\pi i(\nu_0 - \nu) - 1/\tau}.$$

5.5 Useful functions

This gives the spectral *amplitude* of the radiation. The spectral power density, the square modulus of this, is $\overline{A(\nu)}^*\overline{A(\nu)}$, which is

$$I(\nu) = \frac{|A_0|^2}{4\pi^2(\nu_0 - \nu)^2 + 1/\tau^2},$$

and this is the profile of a spectrum line emitted in an atomic dipole transition. The same function can be derived quantum mechanically. It is known as a Lorentz profile.

5.5.4 The Voigt profile

This is a common shape for spectrum lines which have been broadened by temperature in the emitting gas or plasma.

Temperature alone will broaden a monochromatic line to a Gaussian:

$$I(\lambda) = I_0 e^{-(\lambda - \lambda_0)^2/a^2}.$$

The width parameter, a, comes from the Maxwellian distribution of velocities, and $a^2 = 2\lambda^2 kT/mc^2$, where T is the absolute temperature, k is Boltzmann's constant and m the mass of the emitting species.

Substituting values we find that a monochromatic spectrum line becomes a Gaussian with a FWHM $\Delta\lambda$ given by $\Delta\lambda/\lambda = 7.16 \times 10^{-7}\sqrt{T/M}$ where M is the atomic or molecular weight of the emitter.

The observed spectrum line shape is the convolution of this with the Lorentz profile resulting from the finite decay time of the oscillator, the so-called 'natural' width of the line.

5.5.5 Separation of the components of a Voigt profile

Given a spectrum line with a Voigt profile the width parameters of the two components can be separated. Suppose that the width parameter of the Gaussian component is a and that of the Lorentz component is b. Then the Voigt profile is the convolution

$$V(\lambda) = G(\lambda) * V(\lambda) = e^{-\lambda^2/a^2} * 1/(\lambda^2 + b^2).$$

Let w be the Fourier pair of λ. Then $g(w) \rightleftharpoons G(\lambda)$ and $l(w) \rightleftharpoons L(\lambda)$.

The Lorentz profile $L(\lambda)$ is, by the Wiener–Khinchine theorem, the Fourier transform of the autocorrelation of the truncated exponential function which represents the decaying amplitude of a damped oscillator. The truncation is because the decay is deemed to begin at $w = 0$. Then,

$$l(w') = \int_{w'}^{\infty} e^{bw} e^{-(w-w')/b} \, dw$$
$$= b/2 e^{-w'/b}$$

for positive values of w'. Then the Fourier transform of $V(\lambda)$ is, for positive values of w,

$$v(w) = b/2 \, e^{-\pi^2 w^2 a^2} \, e^{-w'/b}.$$

A graph of $\log v(w)$ is a parabola and the values of a and b can be extracted by simple measurement.

5.5.6 The Dirac delta function

This has the property that $\delta(x)$ is zero everywhere except at the origin $x = 0$, and at $x = 0$ is infinite. The important property is

$$\int_{-\infty}^{\infty} \delta(x) \, dx = 1.$$

It may be regarded as the limiting case of a top-hat function as it becomes infinitesimally wide and infinitely high with its area always unity, or as the limiting case of a Gaussian:

$$\lim_{a \to 0} \frac{1}{a\sqrt{\pi}} e^{-x^2/a^2}.$$

In either case it is an 'improper' function which gives angst to mathematicians but is useful to physicists.[2]

5.5.7 The Dirac comb

This is an infinite set of equally spaced δ-functions each separated from its neighbours by a distance a. It is generally denoted by the Cyrillic letter III (Shah) and written $\text{III}_a(x)$:

$$\text{III}_a(x) = \sum_{n=-\infty}^{\infty} \delta(x - na).$$

Its Fourier transform is then

$$\sum_{n=-\infty}^{\infty} \int_{-\infty}^{\infty} \delta(x - na) e^{2\pi i p x} \, dx = \sum_{n=-\infty}^{\infty} e^{2\pi i n p a}, \qquad (5.7)$$

---

[2] For example, it has no upper bound, which violates one of the Dirichlet conditions that define a function to be Fourier-transformable.

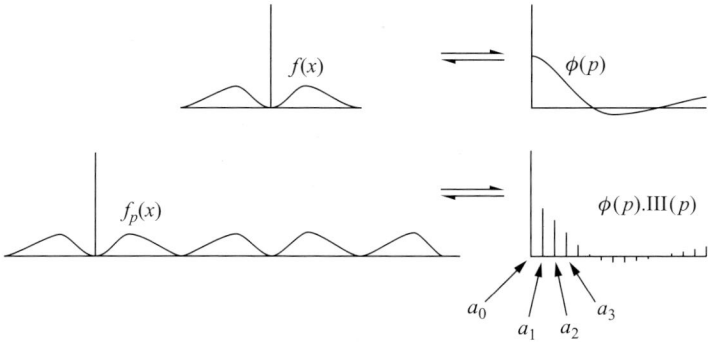

Figure 5.2 The sampling theorem. A continuous truncated function together with its mirror image produces a real Fourier pair. The same function repeated indefinitely can be synthesised from the fundamental a_0 and its 'overtones' $a_1, a_2, a_3, a_4, \ldots$, and these have the same values as the Fourier pair at the appropriate values of p. In practice it is the set of regular samples which provides the experimental database and the function which produced them can be computed via a FFT computer program.

and it can be shown[3] that this is another Dirac comb, $(1/a)\text{III}_{1/a}(p)$, so that we can write

$$\text{III}_a(x) \rightleftharpoons (1/a)\text{III}_{1/a}(x).$$

It is useful for example when describing the aperture of a diffraction grating as the convolution of a top-hat function of width a with a Dirac comb of spacing b, the convoluted product then being multiplied by a wide top-hat function of width Nb to limit the width to N rulings.

5.6 More theorems

5.6.1 The sampling theorem

This is especially important in Fourier–multiplex spectrometry but it applies also to the theory of diffraction grating spectrographs. It comes from Whittaker's interpolary function theory[4] where it is called the *cardinal* theorem, and states, in effect, that the amplitudes of all the frequencies that a signal contains can be measured provided it is measured ('sampled') at intervals of $1/2\nu_f$, that is, to the reciprocal of twice the highest frequency present.[5] It serves also to link Fourier *series* with the Fourier *transform*.

---

[3] R. N. Bracewell, *The Fourier Transform and Its Applications* (New York: McGraw-Hill, 1965).
[4] J. M. Whittaker, *Interpolary Function Theory* (Cambridge University Press, 1935).
[5] Sometimes called the *Nyquist* or *folding* frequency.

The signal which enters a spectrograph or spectrometer has an alternating electric field which can be described by $F(t)$ and this has a spectrum, $S(\nu)$, which may extend from $\nu = 0$ to ν_0.

Consider a periodic function $S_p(\nu)$ which is identical with this spectrum from 0 to ν_0, is the mirror image from ν_0 to $2\nu_0$ and is thereafter periodic with period $2\nu_0$. In other words, it is the convolution of $S(\nu)$ and its mirror image in the origin with a Dirac comb of interval $2\nu_0$.

This periodic function can be written as the sum of an infinite series:

$$S_p(\nu) = \frac{a_0}{2} + \sum_{n=1}^{\infty} a_n \cos \frac{2\pi n \nu}{2\nu_0}$$

$$= \frac{a_0}{2} + \sum_{n=1}^{\infty} a_n \cos \frac{\pi n \nu}{\nu_0}, \qquad (5.8)$$

and then the amplitudes a_n are given by

$$a_n = \frac{1}{\nu_0} \int_0^{\nu_0} S(\nu) \cos \frac{\pi n \nu}{\nu_0} \, d\nu,$$

and the integral here need go only from 0 to ν_0 because the integrand is zero from there on.

Now suppose we have a signal, $F(t)$, which contains frequencies up to ν_0 and is empty from then on. Its spectrum, $S(\nu)$, is continuous from $\nu = 0$ to $\nu = \nu_0$. Corresponding to this spectrum we may construct a symmetrical function $S_s(\nu)$ such that $S_s(\nu) = 1/2(S(\nu) + S(-\nu))$, which is symmetric and stretches between $-\nu_0$ and ν_0. The convolution of this function with $III_{2\nu_0}$ is symmetric and periodic with a period $2\nu_0$ and can therefore be synthesised from a discrete infinite set of cosines with coefficients which are a set of numbers a_n. In Fourier language:

$$S_s(\nu) = \frac{1}{2}[S(\nu) + S(-\nu)] * III_{2\nu_0}$$

and the corresponding $F_s(t)$ is

$$F_s(t) = \frac{1}{2}[F(t) + F^*(t)]III_{\frac{1}{2\nu_0}}(t).$$

Since $F(t)$ is real, $F(t) = F^*(t)$ and $F_p(t) = F(t)III_{\frac{1}{2\nu_0}}(t)$, which is the set, $F(nt_0)$, of 'samples' of $F(t)$ at intervals $t = 1/2\nu_0$. From this set of samples we recover $S_s(\nu)$ by the Fourier sum:

$$S_s(\nu) = \sum_{-\infty}^{\infty} F(nt_0)e^{2\pi i n\nu/\nu_0} = \sum_{-\infty}^{\infty} F(nt_0)e^{\pi i \nu n t_0}.$$

Thus, in principle, a measurement of the incoming signal at intervals $1/2\nu_0$ is sufficient to give the spectrum of the signal. The spectral resolution depends on the number of samples taken of the signal, when the '∞' signs in the sum are replaced by 0 and N, the (finite) number of samples taken.

The application of this theorem is particularly apparent in Fourier spectroscopy, but the analogy in grating spectroscopy is apparent also when considering the sum of amplitudes reflected from a grating. The angle of diffraction and the implied phase delay in the signal are analogous to the exponential in the Fourier sum and different values of ν correspond to different angles.

The detector receives the sum of the complex amplitudes from the rulings, in other words the sum of the samples, each delayed in time from its neighbours by the extra path difference it has travelled. This sum multiplied by its complex conjugate is the power detected at each point of the spectrum.

5.7 Aliasing

The 'highest frequency', ν_0, in the last section is also called the *folding frequency* of the spectrum: the frequency beyond which there should be no power. If the sampling interval is too large, the corresponding folding frequency may be too low for the spectrum and the power beyond it at frequency ν appears in the spectrum at frequency $\nu_0 - \nu$. Power at frequency $2\nu_0$ appears at frequency 0. For example, consider a cosine of amplitude a. It must be sampled at 0, π, 2π, 3π etc., and alternate samples will have values $a, -a, a, -a \ldots$ From this the amplitude is correctly recovered by the summing process. Had the samples been taken at 0, 2π, 4π, $6\pi \ldots$ they would all have had the same value, a, and the cosine would appear to be at zero frequency.

Again, had the samples been taken at 0, 3π, $6\pi \ldots$ the frequency would have appeared in its proper place.

Aliasing can thus be put to good use. If the spectrum is confined to a small part of the frequency domain it can be under-sampled deliberately. It may in fact occupy the third, fifth etc. period of the periodic function, $S_s(\nu)$, but the Fourier-summing process will put it apparently in the first period. This is perfectly satisfactory provided we know beforehand what the sampling interval and the range of the spectrum are, and it avoids the tedium of sampling for the empty part of the spectrum at lower frequencies.

Again the analogy with grating spectra will be clear: we may look at the second or third orders of diffraction, knowing beforehand that the first-order spectrum at longer wavelengths is absent, either by filtering or by detector impotence.

6
Physical optics and diffraction

6.1 Fraunhofer diffraction

Elementary textbooks dealing with a transparent aperture in physical optics reduce the question of Fraunhofer or 'far-field' diffraction to a simple two-dimensional form in which a plane wavefront passes through an opaque surface. There is no variation perpendicular to the plane of the diagram and in this third dimension the aperture is assumed to have unit length. A narrow strip of width dx along the x-direction is a source of 'secondary wavelets' and transmits, according to Huygens' principle, a wave with an amplitude proportional to the area $A\,dx$, where A is the amplitude per unit area on the incident wavefront. In the far-field approximation the inverse-square law is ignored and the amplitude, including the phase,[1] at the point P is dU, where

$$dU = A\,dx\,e^{\frac{2\pi i}{\lambda}(R - x\sin\theta)},$$

where R is large compared with λ and x. If the amplitude A varies from point to point along the aperture, the total amplitude arriving at point P is

$$U(\theta) = e^{\frac{2\pi i}{\lambda}R}\int A(x)\,e^{-\frac{2\pi i}{\lambda}x\sin\theta}\,dx,$$

and the integral is over the whole of the transparent part of the aperture.

If A varies in the x-direction and if we replace $\sin\theta/\lambda$ by a new variable p and take the infinite integral we find

$$U(p) = C\int_{-\infty}^{\infty} A(x)e^{-2\pi i p x}\,dx \qquad (6.1)$$

so that

The amplitude $U(p)$ diffracted in a direction θ is proportional to the Fourier transform of the incident amplitude $A(x)$ across the aperture

---

[1] Remember that phase change = $\frac{2\pi}{\lambda} \times$ path change.

6.1 Fraunhofer diffraction

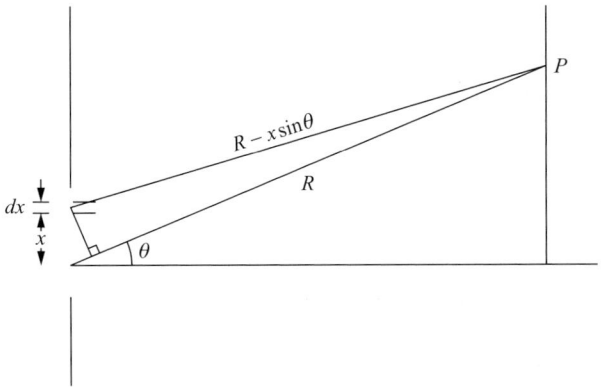

Figure 6.1 Fraunhofer diffraction by a slit aperture. Each element of the aperture is treated as a separate source coherent with all the others and the several amplitudes are added at the image point to produce a resultant which depends on the various phases as well as the magnitudes of the elements.

and as usual in optics, $A(x)$, the variation of incident amplitude with x, when multiplied by its complex conjugate, is the transparency of the aperture.[2]

In practice it is the *intensity* of the diffracted light which we measure and this in turn is the (complex) diffracted amplitude multiplied by its complex conjugate, and we write, remembering that $C = e^{\frac{2\pi i}{\lambda}R}$ so that $CC^* = 1$,

$$I(p) = U(p)U^*(p).$$

This can now be applied to various apertures and the usual diffraction patterns are described.

6.1.1 Single-slit diffraction

A single slit of width w will diffract an amplitude $\bar{a}(p)$ where

$$\begin{aligned}
\bar{a}(p) &= A \int_{-w/2}^{w/2} e^{-2\pi i p x}\, dx \\
&= A \frac{1}{2\pi i p}(e^{\pi i p w} - e^{\pi i p w}) \\
&= A w \frac{1}{\pi p w} \sin(\pi p w) \\
&= A w \operatorname{sinc}(\pi p w),
\end{aligned}$$

where the sinc function ($\sin x / x$) is as described in the previous chapter.

---

[2] $A(x)$ itself may be complex, since there may be variations in the optical thickness in what is otherwise a transparent part of the aperture, and this will cause the phase to vary with x. This happens for instance in the blazed transmission grating.

54   *Physical optics and diffraction*

The diffracted intensity then varies according to

$$I(p) = I(0)\text{sinc}^2(\pi p w),$$

which is the familiar expression for diffraction by a single slit.

6.1.2 Two-slit diffraction

Two slits, as in the Young's slits experiment, are represented by an aperture function which is the convolution $\Pi_w(x) * [\delta(x - a/2) + \delta(x + a/2)]$, where w is the width of each slit and a their separation.

The Fourier transform of this is $2\,\text{sinc}(\pi p w)\cos(2\pi p a)$, and this, when multiplied by its complex conjugate, describes the familiar two-slit diffraction pattern.

Bear in mind that we can rejoin physical reality at any time by replacing the abstract variable p by $\sin\theta/\lambda$.

6.1.3 N-slit diffraction: the diffraction grating aperture

The original crude diffraction gratings were transparent slits ruled on an opaque screen of smoked glass. Such an aperture can be represented by narrow top-hat functions representing a single ruling, a Dirac comb representing the grid of apertures and a broad top-hat function defining the limited width of the grating. The result is

$$G(x) = \Pi_{Na}(x)[\Pi_w(x) * \text{III}_a(x)]. \tag{6.2}$$

The width of a ruling is w, the spacing between adjacent rulings (the grating constant) is a and there are a total of N rulings.

The Fourier transform of this is

$$\Gamma(p) = Na\,\text{sinc}(\pi N p a) * \left[w\,\text{sinc}(\pi p w)\frac{1}{a}\text{III}_{1/a}(p)\right], \tag{6.3}$$

and with $p = \lambda/\sin\theta$ as before, this is the amplitude diffracted in the direction θ. Notice in particular that because of the convolution with a Dirac comb, $\Gamma(p)$ will have a maximum whenever $p = n/a$, that is when $\sin\theta = n\lambda/a$. n is any integer, positive or negative and λ/a is the ratio of wavelength to slit spacing. The application of this to diffraction grating theory will be elaborated in Chapter 8.

6.1.4 Diffraction with oblique incidence

If the wavefronts are incident on the diffracting aperture at an angle ϕ to the aperture normal, we must take account of the phase at the aperture. The complex amplitude

6.2 Two-dimensional apertures and oblique incidence

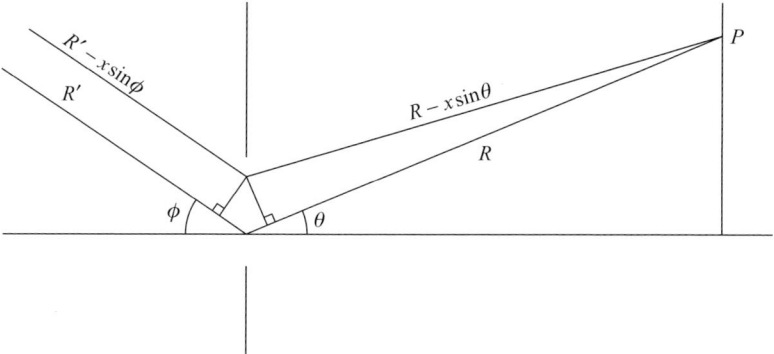

Figure 6.2 Fraunhofer diffraction with oblique incidence. The different times of arrival of the incoming wavefront at the different elements of the aperture imply an extra phase term in the eventual sum of the elementary amplitudes at the image point.

at $z = 0$ is then

$$A(x)e^{\frac{2\pi}{\lambda} x \sin\phi},$$

and the diffracted amplitude is

$$U(p) = \int_{-\infty}^{\infty} A(x) e^{\frac{2\pi}{\lambda} x \sin\phi} e^{\frac{2\pi}{\lambda} x \sin\theta} \, dx.$$

For N slits, where $A(x) = \Pi_{Na}(x)[\Pi_w(x) * III_a(x)]$, the diffracted amplitude has a maximum whenever $\sin\theta + \sin\phi = n\lambda/a$.

6.2 Two-dimensional apertures and oblique incidence

Assume that the diffracting aperture in three-dimensional Cartesian coordinates occupies the plane $z = 0$. A plane wavefront, incident obliquely, is conveniently described by its direction cosines, l, m, and n. If the phase is zero at the origin, the wavefront occupies the plane $lx + my + nz = 0$ and the phase at the point $P(x, y, 0)$ is $(2\pi/\lambda)$ times the perpendicular distance $(lx + my)$ from the point to the diffracting plane. The two-dimensional complex amplitude at the diffracting aperture is then

$$A(x, y)e^{\frac{2\pi i}{\lambda}(lx+my)},$$

and the diffracted amplitude is

$$U(p, q) = \int_{-\infty}^{\infty} A(x, y) e^{\frac{2\pi i}{\lambda}(lx+my)} e^{\frac{2\pi i}{\lambda}(px+qy)} \, dx \, dy,$$

where p and q are, as before, the direction cosines of the diffracted wavefront.

$A(x, y)$ for a diffraction grating is separable into $A_1(x) A_2(y)$ and $A_2(y)$ is a top-hat function with a width equal to the length of the grating rulings. The condition now for a maximum of the diffracted amplitude is obtained as before, when $l + p = n\lambda/a$ and $m + q = 0$, but l and p are now direction *cosines*. $l = \sin i \cos \theta$ and $p = \sin r \cos \theta$, and the equation for a maximum, in Cartesian coordinates, becomes

$$(\sin i + \sin r) \cos \theta = n\lambda/a, \tag{6.4}$$

where θ is the angle of inclination of the incoming chief-ray to the x–z plane.

As with the diffraction grating, the amplitude at the diffracting aperture is often separable into functions of x and y, and the integrals are separated and elementary. However, if there is circular symmetry, the whole integral equation must be referred to polar coordinates. The problem is not usually to be found in spectrograph design and will not be pursued further here.

7
The prism spectrograph

7.1 Introduction

This is the traditional form of spectrograph or spectroscope, first perfected by Bunsen and Kirchhoff. In its transmission form the prism is generally an equilateral triangle in section. There is no absolute virtue in this as it is the prism base length which determines the resolving power, and the equilateral is a compromise between too great an apex angle, which results in more light reflection at the surfaces, and too small an apex angle, which results in an extra weight of glass at no extra resolution. The angle of $60°$ is also close to the Brewster angle[1] for one of the planes of polarisation, which gives – in principle – perfect transmission for that plane and a guaranteed 50% throughput.

7.2 The traditional prism spectrograph

This is still useful for rapid determination of refractive index.

The resolving power is determined by the wavelength range corresponding to the FWHM of the diffraction pattern produced at the image of a slit in monochromatic light. In Fig. 7.1 we have

$$h = s\cos\frac{(A+D)}{2}, \qquad b = 2s\sin\frac{A}{2}, \qquad n = \frac{\sin\left(\frac{(A+D)}{2}\right)}{\sin\left(\frac{A}{2}\right)},$$

whence

$$\frac{\partial n}{\partial \lambda} = \frac{\cos\left(\frac{(A+D)}{2}\right)}{\sin\left(\frac{A}{2}\right)}\frac{1}{2}\frac{\partial D}{\partial \lambda},$$

---

[1] It would be exact for a refractive index of $\sqrt{3} = 1.732\ldots$

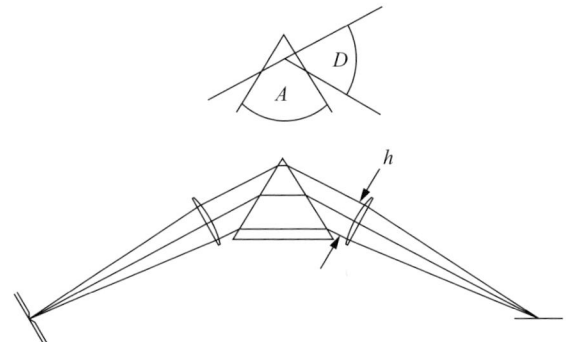

Figure 7.1 The spectrograph derived from the classic prism spectroscope of Kirchhoff and Bunsen. The resolution depends on the width, h, of the emerging beam. The ratio λ/h determines the optical resolution via diffraction theory and the dispersion requires an angle greater than this between two resolved wavelengths.

and on making the necessary substitutions:

$$b\, \partial n/\partial \lambda = h\, \partial D/\partial \lambda.$$

The Rayleigh criterion for resolution is $\delta D = \lambda/h$ in this instance so that

$$\frac{\lambda}{d\lambda} = R = b\frac{\partial n}{\partial \lambda},$$

where b is the base length of the prism. Visual inspection with the telescope on a goniometer table divided into minutes of arc will allow the refractive index to be measured with a precision of 1 part in 10^4, and interpolation between two known lines on a photograph will achieve a wavelength measurement with an accuracy of a few parts in 10^6, bearing in mind that the centre of a diffraction-limited line image can be estimated to $\sim$1%.

$\partial n/\partial \lambda$ tends to vary rapidly with wavelength in highly dispersive glasses and dispersion curves for UV-quality fused quartz and a typical 'short flint' glass suitable for prisms are shown in Fig. 7.2.

One of the chief reasons why prism spectrographs are still worth considering is their freedom from grating 'ghosts'. The line profile, provided that the prism is the limiting aperture, is a sinc^2 function, and may have its secondary maxima further reduced by a mask. This makes it sensitive to faint satellite lines on either side of a parent line and this, together with the relative freedom from scattering when best-quality glass is used, makes it suitable for Raman spectrography, where spectral purity and not high resolution is the major criterion. In a Littrow mounting, where the base of the 30°–60°–90° reflecting prism is 75 mm, the resolving power of a Schott SF11 glass prism at 5000 Å is 40 000, so that a Raman satellite 0.12 Å (0.5 cm^{-1} or 6×10^{-5} eV) from the parent line can in principle be resolved provided it is bright compared with the first satellite of a sinc^4 line shape.

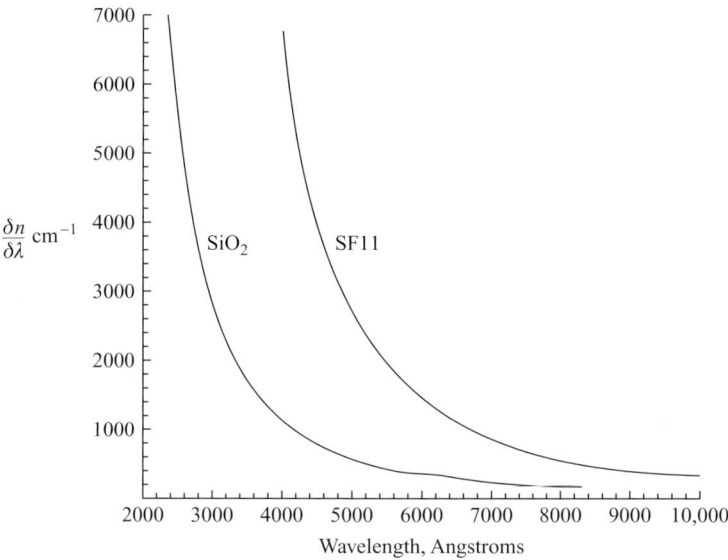

Figure 7.2 The wavelength dispersion curves for two materials suitable for spectrograph prisms. The 'short flint' SF11 is satisfactory at wavelengths above 3200 Å and fused quartz is unsurpassable between 3000 Å and 2000 Å. The curves show resolving power of a prism with a 1 cm base length.

7.3 The focal curve theorem

This derives from the elementary consideration of chromatic aberration in a lens. If most of the lens is removed, the images of a point object will appear in focus along the optic axis, the red image most distant from the lens vertex. By extrapolation, a prism with lens of the same glass type to collimate and focus the light will show the spectrum of a point source lying along a line joining the red image to the source.

This is not exact of course, but it explains why the detector in such a spectrograph lies at a steep angle to the emerging chief-rays. The focal surface is generally not flat but is a gentle curve which must be determined by calculation or ray tracing. Achromatic or apochromatic lenses may be computed for collimator and camera which will give a flat focal surface, and camera lenses can be pressed into service on both sides to give a low focal ratio, but it must be borne in mind that such lenses are computed strictly for the visible region and lose resolution very rapidly outside their intended spectral extent, especially at the UV end of the spectrum.

7.4 The Littrow mounting

This uses a 30°–60°–90° prism with a mirror to reflect the light, giving in effect a 60° prism with half the weight of glass. With a high focal ratio and a singlet lens

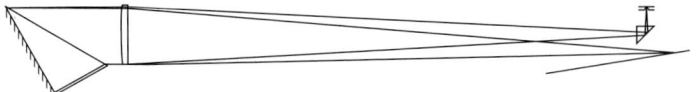

Figure 7.3 The traditional quartz Littrow mounting. With a resolution of about 0.1 Å throughout the visible and near UV, this was state-of-the-art spectrography in the mid twentieth century and much used for the analysis of molecular spectra.

acting as collimator and camera diffraction-limited resolution is possible, although long exposure times are needed, and the photographic plate is at a steep angle to the output chief-rays.

The Hilger $1\frac{1}{2}$-metre quartz Littrow spectrograph (Fig. 7.3), which gave yeoman service in the first half of the twentieth century, was a typical example of a high-performance Littrow prism spectrograph, working at a focal ratio of $\sim$F/30 and giving resolving power in the region of 40 000 in the near UV. The recording plates were 250 mm long by 100 mm wide, were curved gently by pressing them against a mandrel in the plateholder and gave full resolution along the whole length of the spectrum. Ten or more spectra could be recorded on one plate by moving the plateholder vertically with a rack-and-pinion motion.

7.5 The Pellin–Broca prism

This was an ingenious device (Fig. 7.4) for producing a monochromator with constant deviation – in this case 90° – by burying the two halves of the usual prism into one block of glass with a 90°–45°–45° prism acting as a reflector. The wavelength presented at the output slit or eyepiece could be varied by rotating the prism slowly with a turntable and worm-screw.

It is in effect a 60° prism, split into two and with a 45° prism using total internal reflection in place of a mirror. It was a feature of the constant deviation spectroscope, where wavelength scanning was by simple rotation of the prism table rather than the whole of the telescope. Many remain far back in the store rooms of university physics departments.[2] The resolving power is determined by $\partial n/\partial \lambda \times 2 \times$ base length of the smaller of the two buried 30°–60°–90° prisms (Fig. 7.4).

A serviceable high-speed, low-resolution spectrograph can be made from this prism, using photographic lenses for collimation and photography and Fig. 7.5 shows an example using a 200 mm F/3.5 Tessar as collimator and a 90 mm F/1.8 double-Gauss 35 mm camera lens to focus the spectrum on to a CCD. The dispersion gives a spectrum from 4000 Å to 7500 Å of about 11 mm length, depending on the glass type in the prism. With 5 µm pixels in the CCD the resolution is about 1.8 Å.

---

[2] The author's own example has a face length of 105 mm on the larger 30°–60°–90° prism, a thickness of 45 mm, a mass of 886 g and acts as a paperweight on his desk.

7.6 Focal isolation

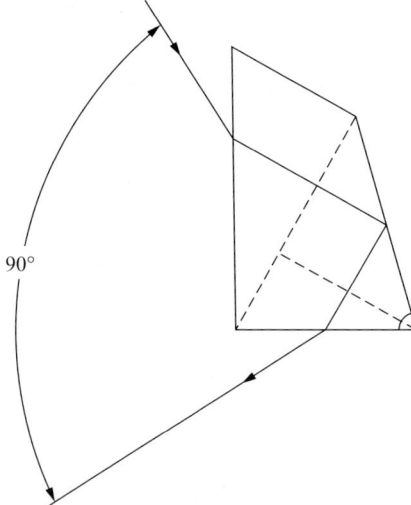

Figure 7.4 The Pellin–Broca prism, which at the 'minimum deviation' position gave an output beam perpendicular to the input beam. This in turn gave rise to the 'constant deviation' spectroscope, which was scanned in wavelength by rotating the prism on a turntable while the collimator and telescope were fixed.

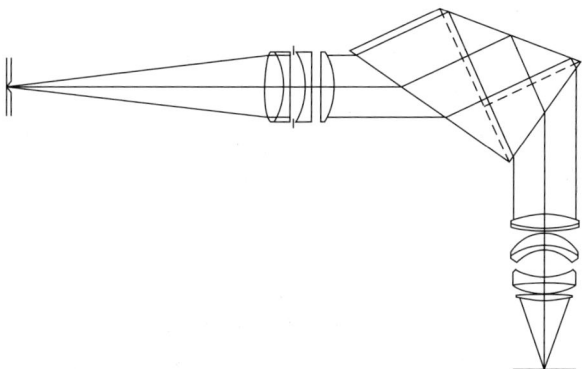

Figure 7.5 The Pellin–Broca prism in a serviceable spectrographic mounting. Standard camera lenses can be used for collimator and camera and in the vicinity of 5000 Å for example will give a resolution in the region of 0.5 Å over a limited range.

7.6 Focal isolation

This is a technique invented by H. Rubens and R. W. Wood initially for investigating 'heat waves'.[3] Essentially it is a prism technique, relying on chromatic aberration for its effect. In its original form it comprised a lens, a source and a screen with a small hole to transmit light through to the detector. With the chromatic aberration

---

[3] R. W. Wood, *Physical Optics* (New York: McMillan, 1911), 414.

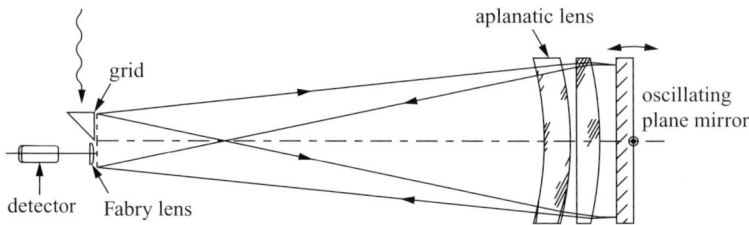

Figure 7.6 The focal isolation principle. A not yet fully explored possibility for infra-red spectrometry when a very high étendue is required and new high-dispersion infra-red transmitting materials may be employed. It has the merit of simplicity and portability. Modulation of the focused wavelength can be by rocking the reflecting mirror or by vibrating the grid in its own plane perpendicular to its rulings.

of the lens, only one small wavelength band is in focus and fully transmitted. They achieved some success in crude spectrophotometry using a Welsbach lamp and observed radiation out to 0.1 mm wavelength.

It was developed considerably, at first into an ultra-violet spectrometer[4] but more lately into a high-resolution infra-red spectrometer[5] by manually figuring a lens with a high infra-red dispersion so as to correct it for spherical aberration. One or more aplanatic elements were added of the same material to improve the relative aperture, and it imaged a fine grid on to itself at strictly −1 magnification. The grid was made to oscillate, mechanically or optically, perpendicular to the grid elements through one half-space and this modulated the wavelength for which the focus was good. It was scanned in wavelength by moving the lens along the optic axis. It had the advantage of a very large étendue, unlimited by the usual constraint of a fixed L-R product.

Its demerits were (1) the line profile, which as a result of false resolution[6] is a sinc function with its huge attendant side-lobes, and (2) it suffered the multiplex disadvantage, whereby radiation is received from all spectral elements but only one is modulated. Consequently, like the Michelson Fourier spectrometer, it is effective only when detector noise is the limiting noise-source.

The device is not yet well developed and more experiment, in modulation methods and especially in lens design to improve the dispersion and line profile while retaining the resolution, is probably needed in order to realise its full potential.

[4] C. S. Forbes, L. J. Heidt & L. W. Spencer, *Phil. Mag.*, **5** (1934), 253.
[5] J. F. James & R. S. Sternberg, *J. de Phys.*, **20** (1967), C2-326; L. McNaughton, Ph.D. thesis, University of Manchester (1969).
[6] H. H. Hopkins, *Proc. Roy. Soc.*, A, **231** (1955), 98.

8
The plane grating spectrograph

8.1 The shape of a monochromatic line spectrum

In Chapter 6 the expression was derived for the diffracted amplitude from a diffraction grating:

$$\Gamma(p) = Na\,\text{sinc}(\pi Npa) * \left[w\,\text{sinc}(\pi pw)\frac{1}{a}\text{III}_{1/a}(p)\right], \qquad (8.1)$$

where a is the interval between rulings and w is the ruling width.

The intensity of light diffracted by a grating is then

$$I(p) = N^2 w^2 \text{sinc}^2(\pi Npa),$$

and this is modulated by the intensity represented in the square bracket, that is, by $\text{sinc}^2(\pi pw)$ repeated at $p = 1/a, 2/a, 3/a, \ldots, n/a$.

The first factor gives the shape of each spectrum line, a very narrow sinc^2 function determined by the total ruled width.

In the second factor (Fig. 8.2b), the first component of the convolution describes the broad sinc function that controls the intensities of the various orders of diffraction, and the second component gives the values of $\sin\theta/\lambda$ at which diffraction maxima occur.

Figure 8.2c shows the complete graph describing the diffracted intensity.

From the shape of the line and using Rayleigh's criterion[1] as a guide, we can deduce the resolving power of the grating. Maxima of the pattern, spectral lines that is, occur when $p = n/a$. Using the sinc^2 function of Fig. 8.2a, the shape of the line is

$$I(\theta) = \text{sinc}^2(\pi Na\sin\theta/\lambda).$$

---

[1] Lord Rayleigh, *Phil. Mag.*, **8** (1879), 261.

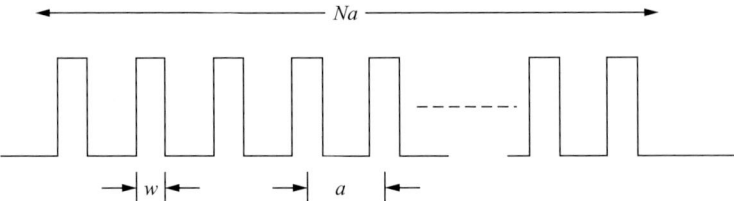

Figure 8.1 The primitive geometry of the grating. The transparent width of each slit is w and the distance between adjacent rulings is a. The overall width of the ruled area is Na and these three parameters are used in the derivation of the grating equation.

Rayleigh's criterion for resolution required a 20% fall in intensity between two maxima. This was doubtless due to the fact that when two identical sinc^2 functions are added, each displaced so that its central maximum lies on the first minimum of the other (Fig. 8.3a), the resulting curve (Fig. 8.3b) has a 20% dip from the two maxima. This follows from the simple fact that the value of sinc$^2(x)$ at $x = \pi/2$ is 0.405.

Apart from anything else, this is mathematically convenient. At a maximum:

$$\pi p N a = n\pi \rightarrow \sin\theta = n\lambda/Na,$$

where n is any integer – the *order* of diffraction. The adjacent zero is when

$$p = n\pi + \pi/N \rightarrow \sin\theta = (n + 1/N)\lambda/a.$$

If this is the same θ as the maximum of wavelength $\lambda + \delta\lambda$ then

$$\sin\theta = (n + 1/N)\lambda/a = n(\lambda + \delta\lambda)/Na,$$

whence

$$\lambda/\delta\lambda = Nn = R = \text{the resolving power}.$$

This is the theoretical maximum resolving power available. Two comments are required.

(1) The imperfections of manufacture make this practically unrealisable. In practice one may achieve ~85–90% of this. Again in practice it is rare for a grating to be worked anywhere near its maximum theoretical performance. For example, a 200 mm ruled width at 1200 rulings/mm implies 240 000 rulings (with a total length of 36 km) and it is rare indeed for a spectrometer to be required to give this resolving power over a wide spectral range.
(2) A spectral feature can be shown to comprise two separate lines even though they are not resolved: the line shape changes, becoming fatter for example, or asymmetrical when adjacent singlet spectrum lines are not.

8.1 The shape of a monochromatic line spectrum

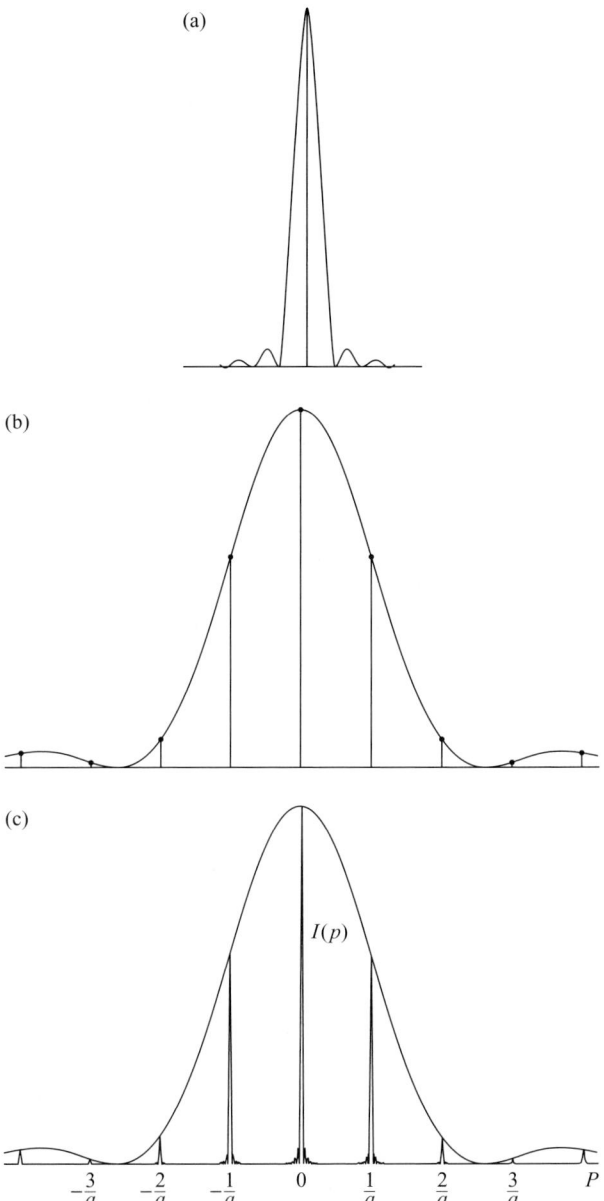

Figure 8.2 (a) The shape of the spectrum line when strictly monochromatic light is diffracted. This is the first factor in the grating equation and the width (FWHM) is determined by the whole ruled width Na of the grating. (b) The second component of the convolution in Eq. (8.1). The amplitudes of teeth of the Dirac comb are controlled by the overall sinc^2 function derived from the width, w, of one ruling. The abscissae here are the sines of the angles of diffraction divided by the wavelength of the diffracted light. (c) When each tooth of the comb is replaced by the sinc^2 function illustrated in (a), the diffraction pattern of the grating is displayed.

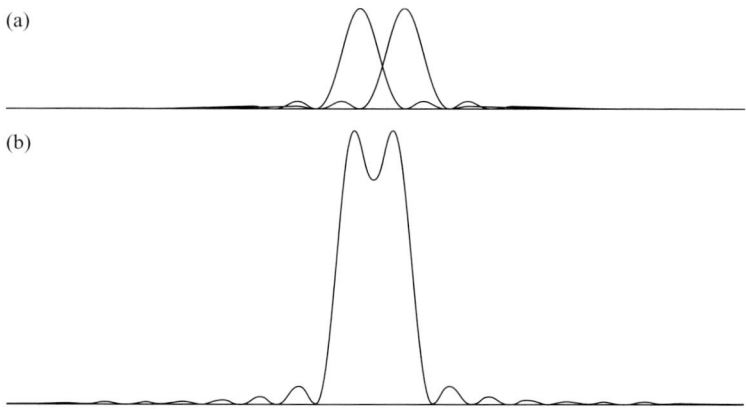

Figure 8.3 (a) Two sinc² functions representing two adjacent spectrum lines lie with one principal maximum lying on the first zero of the other. (b) When the two are added the combined intensity profile has a 20% dip between the principal maxima.

A more elaborate criterion may be taken as a measure of resolution, similar to the optical transfer function used for photographic lenses, in which an image is formed of an object that is a grid of alternate black and white lines. The *visibility*[2] of the fringes in the image, as a function of spatial frequency, is called the contrast transfer function and is regarded as a measure of the spatial filter response of the lens.

In the spectrograph a set of Edser–Butler fringes[3] may be observed, using a Michelson interferometer rather than the usual Fabry–Perot étalon as the fringe generator. With a small diameter source of white light, well collimated as it passes through the interferometer, the intensity at the entry slit of the spectrograph will vary sinusoidally with wavenumber – or approximately with wavelength over a short range of the spectrum. The visibility of the Edser–Butler fringes as a function of path difference in the interferometer is then a measure of the spectral transfer function, which in the absence of aberrations should fall to zero when the Michelson path difference is equal to the ruled width, W, of the grating.

8.1.1 The grating equation

The elementary analysis assumed that light was incident normally on to the grating. If the incidence is oblique, at an angle ϕ to the grating normal, there is an additional phase term[4] covering the whole aperture and then $a = \lambda/(\sin\theta + \sin\phi)$ or, more

---

[2] Defined as $v = (I_{\max} - I_{\min})/(I_{\max} + I_{\min})$.
[3] E. Edser & C. P. Butler, *Phil. Mag.*, **46** (1898), 207.
[4] As in Fig. 6.2.

8.2 Blazing of gratings

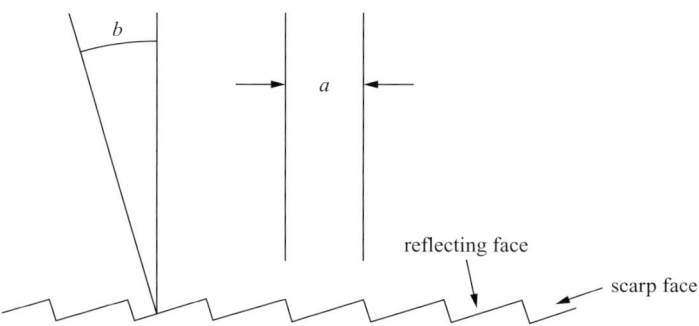

Figure 8.4 A section through the reflecting surface of a blazed grating. The height of the step between rulings, the 'scarp face', is one half-wavelength of the 'blaze' or 'first-order blaze' wavelength. At this wavelength incident light at an angle b to the grating normal is all reflected at this same angle and in first order of diffraction. This is the angle and wavelength of maximum efficiency.

familiarly, the grating equation:

$$\sin\theta + \sin\phi = n\lambda/a. \tag{8.2}$$

Remember that θ and ϕ are on the same side of the grating normal.

8.2 Blazing of gratings

In early gratings where the width of each ruling was about 1/2 the ruling interval, i.e. when w was $a/2$, only odd orders were present, the even orders having been suppressed by the zeros of the broad sinc function lying on top of the diffraction maxima.

In practice the grating rulings have the same width a as the spacing and are inclined at the 'blaze angle' b of the grating, so that a cross-section of the surface has the appearance of Fig. 8.4. The effect of this blazing on a reflection grating can be described by changing in Eq. (6.2) the amplitude factor of one ruling from $\Pi_a(x)$ to $\Pi_a(x)e^{(2\pi/\lambda)2x\sin b}$. The factor of 2 allows for the reflection. Replacing $2\sin b/\lambda$ by q changes the corresponding Fourier transform factor in Eq. (6.3) from $w\,\text{sinc}(\pi pw)$ to $w\,\text{sinc}(\pi pa)*\delta(p-q)$ which, by the shift theorem (Eq. (5.3)) is the same as $w\,\text{sinc}(\pi(p-q)a)$. The maximum of the diffraction pattern then comes when $p = q$, or $\theta = \sim 2b$.

Thus for one wavelength, the *blaze wavelength*, all the incident light is diffracted into first order and the zero order is suppressed as are all other orders, by zeros of the sinc function.

For neighbouring wavelengths the enveloping broad sinc function changes its width at the same time as the diffraction peaks move in the opposite direction so that the maxima are not at the peak transmission. The efficiency is less, and some light is lost into other orders and scattered inside the spectrograph, but nevertheless, as a rule-of-thumb, the efficiency is greater than 50% over a range of wavelengths:

$$\left(\frac{2}{2n+1}\right)\lambda_b < \lambda < \left(\frac{2}{2n-1}\right)\lambda_b,$$

where n is the order of diffraction and λ_b the blaze wavelength.[5]

8.3 Apodising

It is the first factor in the diffraction equation which describes the shape of each spectrum line. A purely monochromatic line is thus presented as a sinc^2 function of FWHM λ/Nn and with secondary maxima on either side of approximately 4% of the height of the parent. It is possible that these secondary maxima may disguise real satellite lines or impersonate fictitious ones, and some effort has been devoted to their suppression. The sinc^2 function occurs when the grating is uniformly illuminated, giving a $\Pi_a(x)$-factor in the aperture function. This factor can be modified by 'masking' the grating with an aperture of a shape such that its Fourier transform alters the shape of the 'line profile'. A diamond-shaped mask for instance will change the first term of the aperture function from $\Pi_{Na}(x)$ to $\Lambda_{Na}(x)$ where the Λ-function, or 'triangle' function, is the convolution of two Π-functions:

$$\Lambda_a(x) = \Pi_{a/2}(x) * \Pi_{a/2}(x).$$

Here, two top-hat functions each of width $Na/2$ are combined to form $\Lambda_{Na}(x)$. By the convolution theorem the Fourier transform of $\Lambda_{Na}(x)$ is

$$(N^2a^2/4)\mathrm{sinc}^2(\pi pNa),$$

and the intensity distribution in a spectrum line is

$$(N^2a^2/4)^2\mathrm{sinc}^4(\pi pNa).$$

This has the merit of suppressing the secondary maxima from $\sim$4.5% to 0.2% of the principal line, a possibly useful gain if the fine structure of a line is being examined. The concomitant demerits are

(1) that by covering half the grating area, half the intensity is lost and
(2) that by restricting the effective width, the resolving power is reduced by $\sim$40%.

There are various shaped masks, known as 'windows', illustrated in Fig. 8.5, which will redistribute the power in the secondary maxima in various ways and the

---

[5] For fine details of the often complicated efficiency curves of diffraction gratings see: C. Palmer, *Diffraction Grating Handbook*, 6th edn. (Rochester, NY: Newport Corporation, 2005).

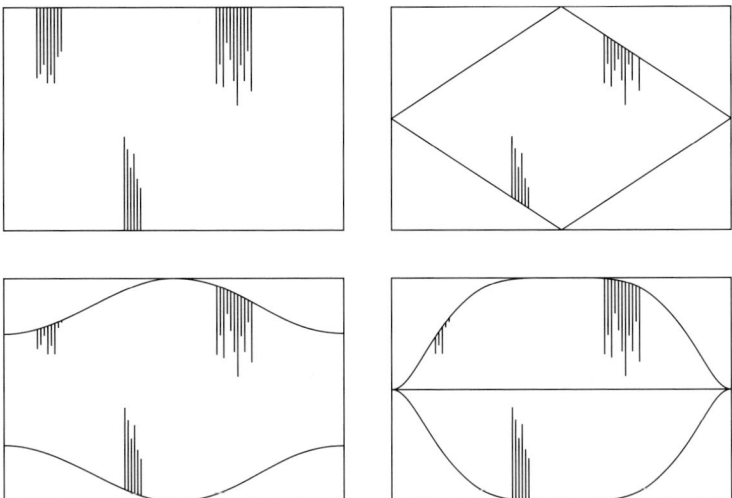

Figure 8.5 Apodising masks for reflection gratings. The 'Hanning' mask with the sinusoidal variation of amplitude across the face of the grating has many advantages, but the area is still open to experiment. The light which fails to reach the principal maximum can be redistributed almost at will among the secondary maxima.

problem of side-lobes is one that is common to all spectroscopy. The 'Hanning' and 'Hamming' windows of communication theory[6] are typical but in the normal spectrograph the combination of the rectangular grating aperture and the usually circular pupils which are superimposed on it compromise all these windows.

The masking process is generally known as 'apodising'[7] and there are legions of apodising masks known and advocated, chiefly by their inventors.

8.4 Order overlap and free spectral range

Despite blazing and other improvements it is not possible to suppress by these means unwanted radiation from other orders of diffraction from appearing in a spectrum. A wavelength λ diffracted in first order is accompanied by $\lambda/2$ in second order, by $\lambda/3$ in third order and so on, to the confusion of the investigator. It may well be that the detector is insensitive to this unwanted radiation but in general precautions must be taken to exclude it. Usually simple colour filtering is adequate and in particular in the visible region in first-order working, a glass which absorbs UV light is sufficient

---

[6] See R. B. Blackman & J. E. Tukey, *The Measurement of Power Spectra* (New York: Dover, 1958).
[7] From the Greek 'without feet'. The same apodising process is used in Fourier-multiplex spectrometry where it is much more effective and more necessary, since the raw instrumental profile there is a sinc function. In that case apodising is part of the computation process and involves interferogram samples being appropriately 'weighted' to form the apodising mask.

to allow an octave of the spectrum – from 8000 Å to 4000 Å – to be photographed unambiguously. However, it may be necessary to use a higher order – with an appropriately higher blaze angle – to obtain the necessary resolution, and in this case narrow-band filters are required to suppress both higher and lower orders.

8.4.1 Free spectral range

If a wavelength λ is to be examined in nth order, then $\sin i + \sin r = n\lambda/a$. At these same angles i and r, $n\lambda$ will appear in first order, $n\lambda/2$ in second order, and generally, $n\lambda/m$ in mth order. To avoid confusion all these other wavelengths must be excluded. The extent of the spectrogram, the range of wavelengths that is, which can be examined without confusion in nth order, stretches from λ to $\lambda - \Delta\lambda$ at which point λ appears again in $(n + 1)$th order. $\Delta\lambda$ comes from

$$(n + 1)\lambda = n(\lambda - \Delta\lambda) \text{ whence } \Delta\lambda = \lambda/n.$$

8.5 Grating ghosts and periodic errors

8.5.1 Rowland ghosts

It sometimes happens that the rulings of a grating are not separated by precisely equal spaces. Particularly in older ruling engines which relied on mechanical precision, a periodic error in the master-screw would cause a cyclic variation in the position of successive rulings. When this happens a monochromatic spectrum line may be accompanied on each side by faint satellite lines known as 'ghosts', sometimes by a whole series of them of steadily decreasing intensity. There is an analogy with the 'side-bands' of a frequency-modulated radio transmission, where the side-bands contain the information signal.

The problem may be described using Fourier theory. The Dirac comb which represented the positions of the rulings is replaced by a set of δ-functions in which the nth is displaced from its proper place by a cyclically varying amount $\alpha \sin(2\pi sna)$. α is the amplitude of the periodic error and $1/s$, the pitch of the lead-screw of the ruling engine, is its period. The aperture function for the grating then replaces the Dirac comb III_a by

$$A(x) = \sum_{n=-\infty}^{\infty} \delta[na + \alpha \sin(2\pi sna)], \quad (8.3)$$

and its Fourier transform is

$$\Gamma(p) = \sum_{n=-\infty}^{\infty} e^{2\pi i p[na + \alpha \sin(2\pi sna)]}, \quad (8.4)$$

with $p = \sin\theta/\lambda$ as usual. From Bessel function theory we have the Jacobi expansion:

$$e^{ix\sin y} = \sum_{m=-\infty}^{\infty} J_m(x) e^{imy},$$

where $J_m(x)$ is the mth-order Bessel function. Using this with $x = 2\pi pn\alpha$ and $y = 2\pi sna$, the sum becomes

$$\Gamma(p) = \sum_{n=-\infty}^{\infty} e^{2\pi ipna} \sum_{m=-\infty}^{\infty} J_m(2\pi p\alpha) e^{2\pi imsna}, \tag{8.5}$$

which, when rearranged is

$$\Gamma(p) = \sum_{n=-\infty}^{\infty} \sum_{m=-\infty}^{\infty} J_m(2\pi p\alpha) e^{2\pi ina(p+ms)}. \tag{8.6}$$

We suppose now that the amplitude, α, of the periodic error is small compared with the grating constant a. For small values of the argument the Bessel functions approximate to

$$J_0(x) = 1, \qquad J_1(x) = \frac{x}{2}, \qquad J_2(x) = \frac{x^2}{4}, \quad \text{etc.}$$

and

$$\Gamma(p) = \sum_{n=-\infty}^{\infty} e^{2\pi inap} + \sum_{n=-\infty}^{\infty} \pi p\alpha \cdot e^{2\pi ian(p+s)} + \sum_{n=-\infty}^{\infty} \pi p\alpha \cdot e^{2\pi ian(p-s)}$$

$$+ \sum_{n=-\infty}^{\infty} \pi^2 p^2 \alpha^2 \cdot e^{2\pi ian(p+2s)} + \sum_{n=-\infty}^{\infty} \pi^2 p^2 \alpha^2 \cdot e^{2\pi ian(p-2s)} + \cdots \tag{8.7}$$

This represents an infinite series of Dirac combs. The first is the one which appears in Eq. (8.1), representing the perfect grating. The second pair represent Dirac combs with amplitudes attenuated by a factor $\pi p\alpha$ and with periods $1/(p+s)$ and $1/(p-s)$. The third pair are attenuated by $\pi^2 p^2 \alpha^2$, with periods $1/(p+2s)$ and $1/(p-2s)$. These describe the 'ghost' lines, and in first order the strongest are separated from the parent line by $\Delta\lambda = \pm\lambda^2/(\text{lead-screw pitch})$. The effect is shown in Fig. 8.6.

8.5.2 Lyman ghosts

These are faint images of a line which appear apparently at sub-integer orders and are chiefly found in XUV spectroscopy when older, mechanically ruled concave gratings are used in first order. They may be ascribed to defects on the ruling tip

72 The plane grating spectrograph

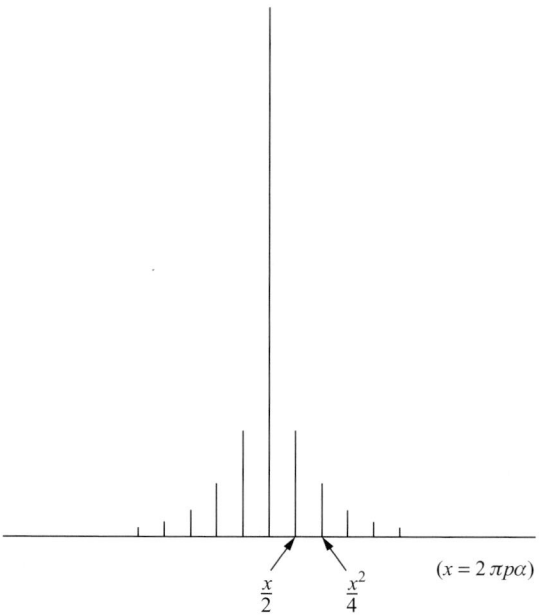

Figure 8.6 Rowland ghosts surrounding a spectrum line. These are separated by amounts which depend on the pitch of the ruling engine lead-screw and by amplitudes which depend on the amplitude of the pitch error. In a modern ruling engine, where the position of the ruling tip is optically controlled, there should be no such periodic error and hence no Rowland ghosts.

which make an irregular but constantly repeated ruling profile so that, perhaps as a result of aliasing, the $III(x)$-function which describes the grating appears to contain another such function in its interstices.[8] This imitates a grating with two, three or more times the number of rulings and consequently two, three or more times the dispersion.

Lyman ghosts are a possible but very unlikely source of confusion unless very large dynamic range and sensitivity are factors in the investigation.

8.6 The complete grating equation

A complete description of the diffraction grating must take account of the length of the rulings as well as the ruled width. The Fraunhofer diffraction pattern is then two dimensional and the two-dimensional Fourier transform must be used with allowance made for incidence which is oblique in both axes.

---

[8] Crudely, if the Π-function which naively describes one ruling were replaced by two narrower Π-functions, the grating would have twice the number of rulings, not necessarily uniformly spaced. If, for example, the Π-functions were separated by $1/3$ of the grating constant, there would be a contribution to the spectrum at order $1/3$.

8.6 The complete grating equation

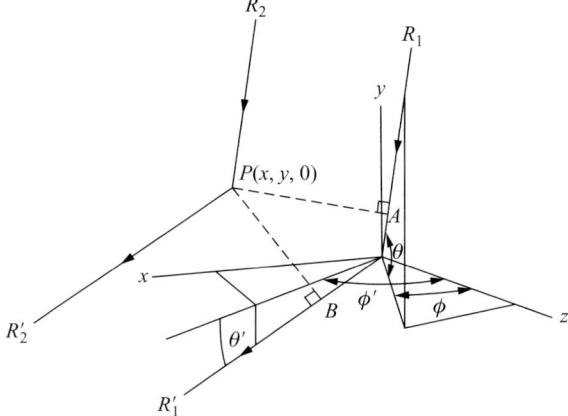

Figure 8.7 The geometry of incident and diffracted rays. R_1 and R_2 are incoming rays from a collimated beam and R'_1 and R'_2 are the same rays after reflection. Perpendiculars from the reflection point of R_2 on to R_1 and R'_1 lie in the incoming and outgoing wavefronts and cut off lengths which add to give the path difference between the two. When the reflection point is one ruling away from the origin the path difference must be one wavelength.

However, the alternative derivation of the full grating equation is both simple and instructive. Suppose that the grating occupies the plane $z = 0$ and that the rulings are parallel to the y-axis. Consider two parallel rays from the collimated incident beam incident on the grating, ray R_1 at the origin and ray R_2 at the point $P(x, y, 0)$ (Fig. 8.7). Suppose that their direction cosines are l_1, m_1, n_1 and that the direction cosines of the corresponding diffracted rays R'_1 and R'_2 are l_2, m_2, n_2. By an elementary theorem in Cartesian geometry, perpendiculars from the point P on to R_1 and R'_1 intercept lengths $PA = l_1 x + m_1 y + 0z$ and $PB = l_2 x + m_2 y + 0z$. This path difference between the two rays must be an integer n number of wavelengths. Thus,

$$AO + BO = (l_1 + l_2)x + (m_1 + m_2)y = n\lambda,$$

where n is the *order of diffraction*.

This must hold for any value of y, so that $(m_1 + m_2) = 0$ and $(l_1 + l_2)x = n\lambda$. If point P is on a ruling adjacent to the y-axis, then the grating equation appears:

$$(l_1 + l_2)x = n\lambda/a, \quad \text{or} \quad (\sin i + \sin r)\cos\theta = n\lambda/a.$$

8.6.1 Distortion of line shapes

Some important consequences come from this equation. Consider first of all the chief-rays of wavelength λ incident on to the grating at its vertex from points on

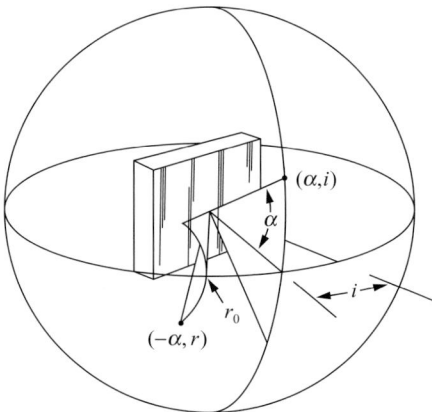

Figure 8.8 The diffraction of rays from different points on a straight entry slit. The complete grating equation requires a conical fan of diffracted rays, which in turn results in a curved image of a straight entry slit.

a straight entry slit parallel to the y-axis. All the x-direction cosines, l_1, are zero. The diffracted chief-rays will therefore all have the same value of l_2, so that they generate a circular cone with its axis on the x-axis of coordinates. If the focal plane of the spectrograph is parallel to the grating surface, the images of the entry slit will occupy a hyperboloid, where the focal plane intersects the chief-rays. Alternatively, if the focal plane is perpendicular to the surface and parallel to the rulings, the images will be arcs of circles. This curvature of line images is potentially a major cause of loss of resolution. The image curvature varies with wavelength and in the calibration of the spectrograph due account must be taken of it.

The proper descriptive tool here is spherical trigonometry, in which the grating normal lies on the equator and the rulings are parallel to the polar axis. In Fig. 8.8, angles like i and r are longitudes and chief-rays approach the centre of the unit sphere from various latitudes and longitudes. A chief-ray coming from latitude α and longitude i is diffracted to latitude $-\alpha$ and longitude r, connected by

$$\sin r = (n\lambda/a)\sec\alpha - \sin i,$$

and for small values of α the approximation $\sec\alpha = 1 + \alpha^2/2$ is adequate. If the equatorial chief-ray is diffracted to r_0, the inclined one arrives at longitude r where

$$\sin r - \sin r_0 = (n\lambda/2a)\alpha^2,$$
$$\delta r = (n\lambda/2a)\alpha^2 \sec r_0. \tag{8.8}$$

For small α the curve of α against δr is a parabola and describes the shape of the Gaussian image of a straight input slit. In practice this may sometimes be

8.6 The complete grating equation

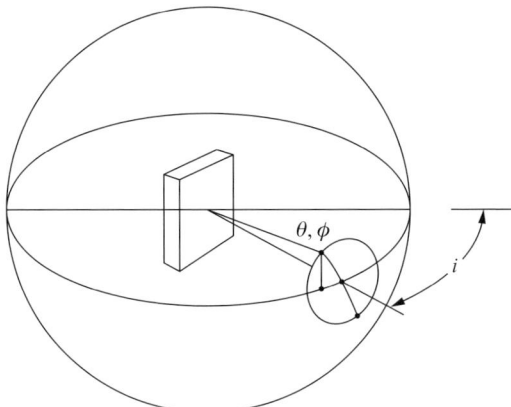

Figure 8.9 An entry slit which is the arc of a circle will produce an image at one wavelength which is a continuation of the same circle. In a scanning monochromator, images of one half-circle will appear on the other half-circle at whatever wavelength is returned from the tilted grating. The wavelength scanning is then done simply by rotating the grating about an axis parallel to the rulings.

compromised by the aberrations of the system but is nevertheless a useful guide to the correction process. The constant of the parabola changes with wavelength and at high resolution each wavelength requires separate treatment to straighten its line image.

8.6.2 The circular slit theorem

Another theorem, chiefly of interest to monochromator constructors, is worthy of mention in this context.

In Fig. 8.9, consider a point at latitude 0, longitude 0 and suppose that this makes an angle i with the optic axis of the spectrometer. Draw a small circle of angular radius ω with its centre at this point. The coordinates (θ, ϕ) of a point on this circle are connected by another well-known theorem in spherical trigonometry:

$$\cos\theta \cos\phi = \cos\omega.$$

Two diametrically opposite points on the circumference of the small circle then have coordinates (θ, ϕ) and $(-\theta, -\phi)$, and if they are also points where incoming and outgoing chief-rays cut the sphere, then

$$\cos\theta[\sin(i+\phi) + \sin(i-\phi)] = 2\cos\theta \sin i \cos\phi = m\lambda/a,$$

or

$$2\cos\omega \sin i = m\lambda/a.$$

Since i is constant, every diametrically opposite pair of points on the small circle are conjugate points for the same wavelength. An input slit which is the arc of a circle will result in an image at wavelength λ on another arc of the same circle. Different wavelengths will provide images at the same circular output slit as the grating is turned about its axis.

8.7 Differential dispersion

In normal practice the spectrograph slit widths are not determined by the ultimate resolving power theoretically available. The entry slit width in such an instance would be such that the first zeros of its diffraction pattern – the sinc^2 function – would fall at the edges of the grating. However, in a spectrograph the resolution may be set by the width of the CCD pixels or the average size of the silver halide crystals of a photographic emulsion. The slit width is then chosen to match this width and there is no resolution to gain and much light to lose in making it narrower. From the grating equation

$$(\sin i + \sin r) = n\lambda/a$$

we get

$$\cos i \, \mathrm{d}i = n \, \mathrm{d}\lambda/a,$$

which relates the resolution dλ to the angular slit width di and for a fixed wavelength,

$$\cos i \, \mathrm{d}i = -\cos r \, \mathrm{d}r, \tag{8.9}$$

which relates the width of the input slit to the width of its image. This is the *lateral magnification*, which is a function of wavelength, and must be taken into account during the design.

8.8 Mounting configurations

Since the time of Bunsen and Kirchhoff and until the end of the nineteenth century the optical arrangement for a spectroscope had comprised a slit, a collimator, a disperser, and a telescope. For photography the eyepiece could be replaced by a photographic emulsion. It was Ebert, following Rowland's production of research-quality diffraction gratings, who proposed an achromatic spectrograph using a single spherical mirror both as collimator and telescope and eliminating the unwanted longitudinal dispersion inherent in a simple lens system.[9]

---

[9] H. Ebert, *Annalen der Physik*, **38** (1889), 489.

8.8 Mounting configurations

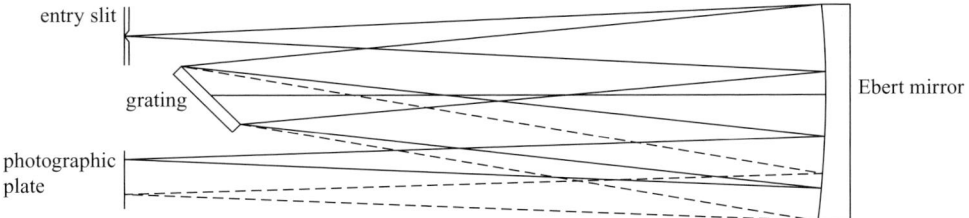

Figure 8.10 The Ebert–Fastie mounting. The first achromatic method of mounting a high-resolution spectrograph, with almost miraculous correction of the gross aberrations. However, it proved to be fiendishly difficult to achieve good optical alignment and it consequently lost ground to the Čzerny–Turner mounting with similar achromatic properties, where the separate mirrors gave more degrees of freedom and greater flexibility in practical alignment.

Strangely it seems to have been ignored until it was rediscovered by Fastie,[10] and meanwhile Čzerny and Turner had replaced the two traditional lenses by two spherical mirrors.[11] Littrow similarly had evolved an arrangement using a single spherical or paraboloidal mirror which would yield diffraction-limited resolution.

8.8.1 The Ebert–Fastie mounting

One of the features of this mounting (Fig. 8.10), which it shares with the Čzerny–Turner mounting, is the partial correction of tangential coma, leaving spherical aberration and astigmatism as the two main aberrations. A proper choice of position for the grating ensures a flat meridional field. Astigmatism is tolerated because only the meridional focus is required and distortion is ignored since the spectral dispersion is in any case non-linear. Although it is capable of yielding a high resolution at a fairly high numerical aperture, the restriction imposed by the single mirror makes the optical alignment of the Ebert mounting unnecessarily difficult and it has largely been discarded in favour of the Čzerny–Turner mounting in most commercial designs. Only two portions of the mirror are used – there can be no overlap between incoming and outgoing beams – and a mask must be inserted in the centre of the aperture to prevent undispersed light falling on to the detector. This reduces the device essentially to the same configuration as the Čzerny–Turner mounting and the Ebert–Fastie arrangement will not be considered further.

---

[10] W. G. Fastie, *J. Opt. Soc. Am.*, **42** (1952), 641 and 647.
[11] M. Čzerny & A. F. Turner, *Z. Phys.*, **61** (1930), 792.

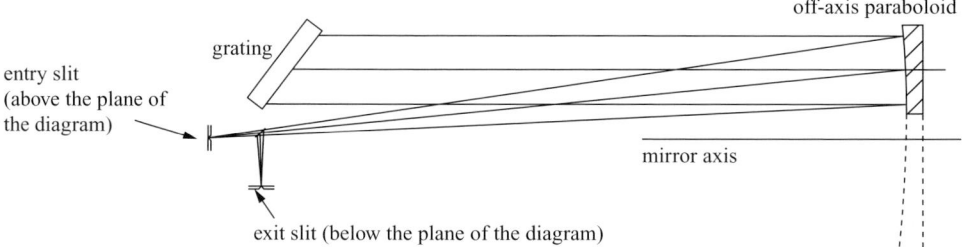

Figure 8.11 The Littrow mounting. Like its prism counterpart it returns the spectrum through the same optical path as the incident radiation and focuses the light close to the entry slit. This version may be constructed with the grating rulings parallel to the diagram, or perpendicular – in which case the grating is rotated instead of tilted as shown here. There is no appreciable difference in the aberration amplitudes or in the spectral resolution available.

8.8.2 The Littrow mounting

This is illustrated in Fig. 8.11. The chief application of this mounting is as a monochromator to give an aberration-free image of a short straight entry slit, and for this reason an off-axis paraboloidal mirror is employed. To keep the field angle as small as possible an 'over-and-under' slit arrangement may be used, with an exit slit for a monochromator and a photographic plate for a spectrograph. As the full coma of a paraboloid is present, the mounting offers no advantage over the Čzerny–Turner mounting as a spectrograph. As a monochromator, high resolution is available provided the entry slit is short[12] and there is the curvature of a line image which follows from the full grating equation.

As a monochromator, the performance is the same when the grating rulings are parallel to the plane of Fig. 8.11 and the grating is turned instead of tilted. The dispersion is then perpendicular to the plane of the figure and the spectrum is formed next to the entry slit. There may be some structural advantage in such an arrangement but in neither instance does the mounting have any particular merit as a spectrograph when compared with the Čzerny–Turner mounting. A major demerit is the need for an off-axis paraboloid, which in turn implies an inordinately large and heavy mirror or one which has been cut to shape from such a mirror. Then there exists the possibility of distortion of the surface figure because of this cutting process, with a consequent loss of resolution. This is a point for discussion with the mirror manufacturer.

An unusual application for the Littrow mounting is in the monochromatic imaging of an extended object, something which is generally done with interference filters. Because the system is virtually stigmatic close to the paraboloid axis,

---

[12] Two per cent of the focal length is usual.

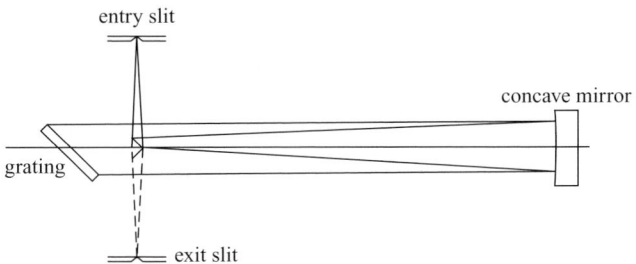

Figure 8.12 The Pfundt mounting. It enjoyed some success as a monochromator mounting but an extended field requires a large prism and consequently an inconveniently large obstruction to the collimated light.

two-dimensional resolution is possible and a rotating mirror may be used to sweep an image of a luminous object – such as the Sun – past the entry slit while a similar mirror – or another part of the same mirror – reflects the output slit synchronously across the field of a camera or telescope. Various auxiliary optical arrangements using mirrors or rotating glass cube prisms are possible and permit the two-dimensional resolution of an object at any wavelength of choice. A ray trace indicates that a spectral resolution of ∼0.05 Å is achievable with a 2 m focal length, a 75 mm ruling width of grating and 15 μm slit widths.

A similar adaptation will convert a Littrow or a Čzerny–Turner monochromator into a spectrograph with a spectral range of a few angstroms. A rotating square-section prism as long as the exit slit and parallel to it can be placed just before the exit slit (i.e. inside the spectrometer) with an electric motor to spin it rapidly about its axis. This sweeps an image of the entry slit across the exit slit and reveals the local structure of a monochromatic emission line multiplet. The detector output swept synchronously across a CRT monitor will show real-time changes to the line structure when, for example, the source temperature is changing or when a magnetic field is applied to it. Self-absorption of the mercury green line as the source warms up is easily demonstrated with this technique, for example.

8.8.3 The Pfundt mounting

This is basically a variant of the Ebert–Fastie mounting where the off-axis aberrations are reduced by using Newtonian mirrors for inserting and extracting the light. It is illustrated in Fig. 8.12. It is of little practical use as a spectrograph because of the narrow field and consequently low spectral range available, but it is a possible alternative to the Littrow mounting when low astigmatism is important in line images. However, the astigmatism does not vanish completely and its chief virtue is its small mirror diameter and consequent low weight, not to mention cost.

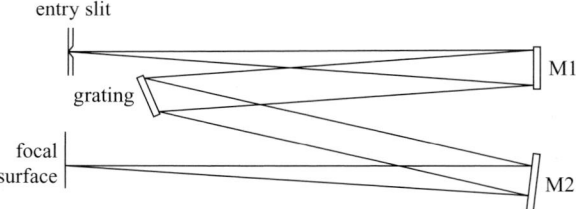

Figure 8.13 The basic Čzerny–Turner mounting. It is often constructed with a symmetrical layout, the grating vertex being at the midpoint between the incoming and outgoing chief-rays, but proper coma correction, according to the Rosendahl criterion, requires sometimes a markedly asymmetric arrangement.

Clearly it can be used with Čzerny–Turner mirrors and then has zero astigmatism. It should probably be called the 'Newtonian' mounting and, like the Littrow, is stigmatic. With paraboloidal mirrors it also has zero spherical aberration. The practical limit, as always, is the curved image of a straight slit which it gives. A possible minor demerit is the necessary obstruction of part of the grating by the Newtonian mirrors with the consequent loss of étendue.

8.8.4 The Čzerny–Turner mounting

This mounting, illustrated in Fig. 8.13, is by far the most popular arrangement for plane reflection grating spectrographs. The two mirrors are used with off-axis chief-rays and there is consequently some coma in the incoming beam. If there were a plane mirror in the grating position reflecting the light to the second mirror, the coma would be exactly corrected, leaving only astigmatism and spherical aberration in the final converging beam. However, after diffraction the outgoing collimated beam is either wider or (more correctly) narrower than the incoming beam, so that if the system is made symmetrical – as many early spectrographs were – the coma is only partially corrected. The result nevertheless at focal ratios of F/12–F/16 gave satisfactory resolution for most purposes.

In the Čzerny–Turner spectrograph the grating tilt should normally be towards the input side. This is an elementary precaution because the output beams are then narrower and allow the greatest length of spectrum to be focused from a given output (M2) mirror diameter.[13] Various alternative arrangements, such as the 'over-and-under' mounting, where the slit, ruling and dispersion directions are perpendicular to the plane of the optics, have been considered from time to time both in Ebert–Fastie and Čzerny–Turner configurations, but they generally confer

---

[13] Or conversely, allow the smallest possible mirror to collect light without vignetting from a given spectral bandwidth.

no obvious advantage, and ray tracing will quickly reveal the huge extra aberration and varying line-tilt incurred.

With the standard layout there are more free design parameters available than are necessary to ensure best performance, and a general rule in design is to keep the off-axis angles as small as possible consistent with the length of spectrum that is to be recorded, and to make the input and output chief-rays parallel. This latter condition greatly simplifies the practical detail of design and construction of the instrument. A note of caution is needed at this point. There must be enough margin between the various ray bundles to allow baffles to be included. Attempts to save space by folding the optical paths with plane mirrors may compromise this feature and it is too important to be neglected or dismissed. For this same reason the so-called 'crossed Čzerny–Turner' mounting is to be regarded with suspicion.

8.8.5 The Rosendahl condition

There was a major improvement in optical design when Rosendahl analysed the aberrations of the Čzerny–Turner configuration and showed that, provided a certain condition, the 'Rosendahl cos-cubed condition' were fulfilled, the tangential coma of the system would be completely corrected at one wavelength.[14] The condition is that

$$\frac{\cos^3 i}{\cos^3 \alpha} \sin \alpha = \frac{\cos^3 r}{\cos^3 \beta} \sin \beta, \qquad (8.10)$$

where i and r are as usual the incident and diffraction angles, and α and β are the field angles of the incoming and outgoing beams at the M1 and M2 mirrors. The condition can be derived from elementary Seidel theory as follows.

In Fig. 8.14, parallel rays starting from the grating and reflected from the M1 mirror will converge to a comatic image at the entry slit, the marginal rays showing an angular coma δ_{coma}. Conversely, marginal rays starting precisely from the entry slit will arrive at the grating edge at this same angle δ_{coma} to the paraxial ray. After diffraction this angle will be multiplied by the factor $(-\cos i/\cos r)$ of Eq. (8.7) and this must be added to the coma produced by the output mirror M2. The total tangential coma is then

$$C_m = -3\mathbf{F_1} y_1^2 \tan \alpha (\cos i / \cos r) + 3\mathbf{F_2} y_2^2 \tan \beta, \qquad (8.11)$$

where subscripts refer to the respective mirrors.

The coma coefficient $\mathbf{F}$ (Eq. 4.8) is given by

$$\mathbf{F} = f^2/4 - \mathbf{Z}(1 - e^2)/8f^2.$$

[14] G. R. Rosendahl, *J. Opt. Soc. Am.*, **52** (1962), 412.

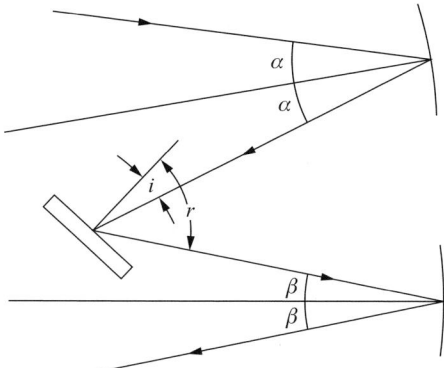

Figure 8.14 The angles involved in the calculation of the coma of the Čzerny–Turner mounting. The rays here are of course the chief-rays of each bundle.

For paraboloidal mirrors, which as we shall see later are preferable to spherical mirrors, this reduces to $f^2/4$.

The geometry of the Čzerny–Turner arrangement provides that $y_1 = (W/2)(\cos i/\cos \alpha)$ and $y_2 = (W/2)(\cos r/\cos \beta)$ and if as usual (though not necessarily) both mirrors are identical, then $\mathbf{F_1} = \mathbf{F_2} = \mathbf{F}$ and the total tangential coma is

$$C_{\mathrm{m}} = -(3/4)\mathbf{F}[(\cos i/\cos \alpha)^2(\cos i/\cos r)\tan \alpha - (\cos r/\cos \beta)^2 \tan \beta], \quad (8.12)$$

and when $C_{\mathrm{m}} = 0$ this is equivalent to the Rosendahl condition.

The condition for vanishing of tangential coma derived here is more general than the Rosendahl condition as it allows for M1 and M2 mirrors of different focal lengths and it also allows exact computation of the coma when there is some offence against the Rosendahl condition. The condition holds strictly only for one wavelength and it is only the tangential coma which is corrected. Sagittal coma remains, albeit somewhat reduced. There is no foreshortening of the aperture on diffraction and the sagittal coma, one-third of the tangential coma, is $\mathbf{F}(L/2)^2 \tan \alpha$. This is converted to $\mathbf{F}(L/2)^2 \tan \alpha(\cos i/\cos r)$ on diffraction, and supplemented by $\mathbf{F}(L/2)^2 \tan \beta$ from the M2 mirror. When the Rosendahl condition is met this gives

$$C_{\mathrm{s}} = \mathbf{F}(L/2)^2 \tan \beta \left[1 - (\sin \alpha/\sin \beta)^{2/3}\right], \quad (8.13)$$

which is substantially less than the coma of a paraboloid at the same field angle away from the zero-coma wavelength.

8.8.6 The generalised Rosendahl condition

To satisfy the Rosendahl condition we abandon the cherished concept of the symmetrical Čzerny–Turner configuration and allow the grating to be nearer (usually) to the input-side optic axis. If we also make allowance for the M1 and M2 mirrors to be of different focal lengths, the condition can be rearranged in a form more convenient for computation:

$$\left(\frac{\cos i}{\cos r}\right)^3 = \left(\frac{f_1}{f_2}\right)^2 \left(\frac{\cos \alpha}{\cos \beta}\right)^3 \left(\frac{\sin \beta}{\sin \alpha}\right). \tag{8.14}$$

Although it is not necessary for the input and output chief-rays to be parallel, it is convenient and it gives us another constraint:

$$r - i = 2(\alpha + \beta) = Q. \tag{8.15}$$

With r replaced by $Q + i$, the grating equation becomes

$$2\cos(i + Q)/2 \cdot \sin(i - Q)/2 = \lambda/a, \tag{8.16}$$

where λ is the desired zero tangential coma wavelength. From Eq. (8.14), i is found by iteration. From Eq. (8.13), r is found. Then by substituting values in Eq. (8.12), α is obtained to any degree of accuracy by another iteration. Finally the grating tilt, t, is given by $t = \pi/2 - 2\alpha - i$.

8.8.7 The flat-field condition

While photographic plates and films can be bent gently to fit the field curvature of an optical instrument, the crystalline nature of a CCD or similar detector requires an accurately flat focal surface. The expression for tangential field curvature of a convex mirror is (Eq. 4.12)

$$C_t = \frac{2}{f} + \frac{3Z}{f^2} + \frac{3}{4}\frac{Z^2(1-e^2)}{f^3},$$

and for a spherical mirror C_t is zero when $Z = -2f(1 \pm 1/\sqrt{3})$, which in practice means that $Z = -0.843 f_2$, where f_2 is the focal length of the output M2 mirror.

Similarly, for a paraboloid the field is flat when $Z = -2f/3$.

In each case, the pupil of the mirror is the grating aperture, and these expressions give the proper dispositions of the M2 mirror following the grating.

The computation does not end here. We must consider the possibility that the optical train will continue beyond the first focus, in which case it is desirable that the output chief-ray at the centre of the field be perpendicular to the focal plane. The Čzerny–Turner configuration is a decentred system where the pupil (the grating) is displaced laterally from the optic axis of the M2 mirror. The problem of the Seidel aberrations of a decentred system is resolved by considering two centred pupils of different diameters and tracing the two marginal rays from one side at different field angles. Their intersection points mark the focal curve for a decentred

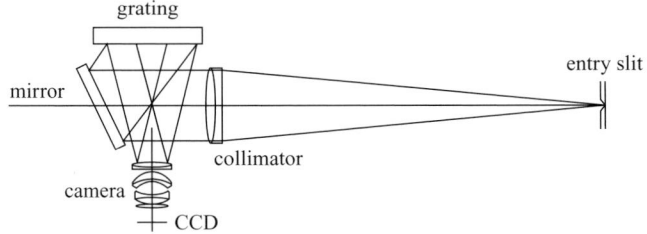

Figure 8.15 The first of two simple mountings which use an oversized grating and make the camera pupil the controlling aperture of the system. Its chief merit is the compact design and the use of standard optical components, thus requiring little design effort or exceptional expense.

pupil. It can be shown that the effect of a decentred stop is to tilt the focal surface about the paraxial focus, and that for a paraboloid the tilt is exactly the tilt of the central chief-ray.[15] As remarked then, this is a fortunate result and means that with a paraboloidal M2 mirror the focal plane can be re-imaged and demagnified with a concomitant useful increase in numerical aperture without loss of resolution. The details of this are in Chapter 13.

8.8.8 Alternative spectrographic mountings

In general the grating is the iris of the system and it is the grating constant which chiefly determines the system's performance. There are simple alternative configurations which the author has found suitable for the examination of very feeble sources at moderate ($R = 1000$–5000) resolution and high numerical aperture, and which are radical departures from the usual rules for spectrograph design. In this case an oversize grating is used and it is the camera iris which determines the performance of the system. The optical diagrams are shown in outline in Figs. 8.15 and 8.16. The important design parameters are the grating constant, its position and its tilt. In each case the chief-ray of the centre of the spectrum leaves the grating close to the normal to the surface and the grating blaze must be towards the input mirror. The wavelength falling on a pixel is proportional to $\sin r$: i.e. $\sin r = \lambda/a - \sin i$ and i is fixed. Thus $\sin r = (\lambda - \lambda_c)/a$ where λ_c is the wavelength at the centre of the spectrum. Alternatively, $\lambda = \lambda_c + a \sin r$.

The spectral range available is set generally by the focal length of the camera lens, the width of the detector and its number of pixels. The position of the camera pupil is determined by the need for the main mirror to image this pupil on to the distant light source – the night sky for example – so that the camera pupil should be at the rear focus of the mirror. The coma and astigmatism of the mirror are

[15] J. F. James, *J. Mod. Opt.*, **41** (1994), 2033.

8.8 Mounting configurations

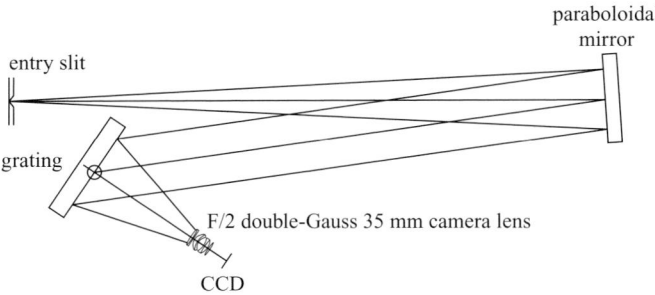

Figure 8.16 This is an alternative simple mounting using an astronomer's 120 mm paraboloid as the collimator. Because of the exceptionally high input-side focal ratio, the coma resulting from the off-axis use is quite negligible. A mottled extended source such as the night sky must first be imaged on to the entry slit to ensure that the grating is uniformly illuminated across its width: otherwise the CCD receives light at different wavelengths from different parts of the source, making spectrophotometry unreliable.

negligibly small, given the low numerical aperture on the input side, and the quality of a 35 mm camera lens is generally sufficient to image the entry slit on to the detector pixels.

8.8.9 Transmission gratings

These are plane gratings with dimensions and grating constants similar to those of reflection gratings. They are transparent mouldings from a master, mounted on a silica blank and have a refractive index, n, in the region of 1.6. They are fully transparent, with inclined grooved rulings which act as a blaze. However, there is refraction as well as diffraction, with attendant total internal reflection at some wavelengths, and this limits their use in spectrography. They are at their best when the wavelength range to be studied is known and fixed and the whole instrument is constructed without moving parts.

They are generally used as in Fig. 8.17 with light incident normally on the plane face and with the camera inclined at the blaze angle. If the grooved faces are inclined at an angle b to the surface, then with normal incidence there will be refraction (neglecting diffraction for the moment) so that the light emerges at an angle r to the normal where

$$\cos r = n - \frac{\lambda}{a \tan b},$$

where a is the grating constant.[16]

---

[16] This refers to the peak of the broad sinc function coming from diffraction by a single ruling, as shifted by refraction.

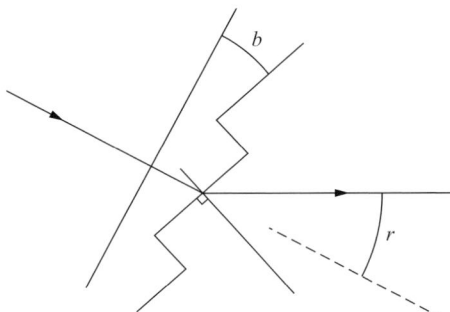

Figure 8.17 The geometry of the transmission grating, illustrating the various angles involved in the text.

This places a limit on the blaze angle allowed. Beyond the limit the refraction causes total internal reflection at the inclined surface of the ruling, and just below it the angle of emergence is so close to the surface that little light is collected by the camera: the camera in fact is looking at the scarp faces of the rulings, from which no diffracted light emerges.

The theory is set by two equations:

$$n \sin b = \sin(b+r) \qquad \text{refraction,}$$
$$\sin r = \lambda/a \qquad \text{diffraction,}$$

where b is the inclination of the ruling surface, r the blaze angle of diffraction and a the grating constant. After some elementary algebra we find the grating geometry set by

$$\lambda/a = (n/2) \sin 2b - \sin b \sqrt{1 - n^2 \sin^2 b}. \tag{8.17}$$

At the same time, the grating ruled width, W_{grat}, is foreshortened and the width of ruling available to the camera is W_{cam}, given by

$$W_{\text{cam}} = W_{\text{grat}} \cos b \cos(b+r). \tag{8.18}$$

The camera also sees scattered light which has been totally internally reflected from the scarp faces of the rulings.

The limitations of performance are shown in Table 8.1. The first column shows the inclination angle of the rulings to the grating surface. The second is the refraction angle of light from one ruling, independent of wavelength and assuming a refractive index, $n = 1.6$. The third is the ratio of the blaze wavelength to the grating constant, a. The fourth is the ratio of the transparent width seen by the camera to the actual ruled width of the grating and the fifth is the corresponding blaze wavelength on the assumption that the grating has 6000 rulings/cm and that the light is normally incident on the plane face.

8.8 Mounting configurations

Table 8.1

Ruling angle	Refraction angle	λ/a	$W_{\text{cam}}/W_{\text{grat}}$	λ_{blaze}
38.5	46.389	0.724	0.070	1.117 μm
35	31.595	0.524	0.325	8131 Å
30	23.130	0.393	0.520	6546 Å
25	17.547	0.301	0.668	5024 Å
20	13.117	0.228	0.787	3799 Å
15	9.463	0.164	0.879	2740 Å

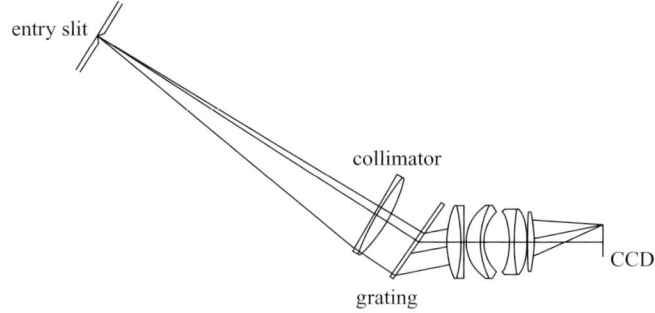

Figure 8.18 A spectrograph with a transmission grating, showing the passage of an oblique ray bundle. Of necessity the camera iris is the defining stop of the system, just as in the spectrographs of Figs. 8.15 and 8.16, and the grating must be proportionately oversized.

The moral of the tale is that no great dispersion is possible because for a long blaze wavelength the grating constant must be large. The broad sinc^2 function caused by the individual rulings ensures that all the diffracted light is either in first order or zero order. The transmission grating is therefore coarsely ruled, typically 3000–6000 rulings/cm, and suitable for fairly low resolution spectrographs. Inevitably, longer wavelengths, being diffracted at larger angles, will suffer a reduction in intensity because of the ruling foreshortening, and shorter wavelengths likewise are attenuated, because the broad maximum mentioned above shifts to larger angles as the refractive index increases while the diffraction angle decreases.

8.8.10 Holographic gratings

These are made by photographing localised interference fringes falling on a photoresist material, followed by chemical or physical removal of those parts unaffected by light or alternatively of those parts which are affected by light. The use of a photoresist is essential in order to get the fine detail required, the reason being that the photosensitivity is at the molecular-bond level rather than the comparatively crude crystal-size level of silver halide emulsions.

The result is a sinusoidal grating in which the grooves follow the lines of the interference fringes. If for example the fringes come from interfering monochromatic 'point' sources the resulting grating, being a hologram, will be stigmatic *for that wavelength*.[17] The groove shape however cannot be controlled to give a blaze as in a ruled grating and the effect of the non-parallel, non-rectilinear rulings on the normal aberrations of a grating spectrometer as well as on its efficiency are open to question. Before deciding to incorporate a holographic grating in a spectrometer, one of the books dealing specifically with the subject must be consulted.

[17] And for adjacent wavelengths too, presumably.

9
The concave grating spectrograph

9.1 The Rowland grating

This is a grating ruled on a concave spherical mirror, in consequence of which it has a self-focusing property and needs no other optical component to produce a spectrum from a point or short line source. It is the invention of H. A. Rowland of Johns Hopkins University who gave an account of it in 1883.[1] The rulings are the intersections of the sphere with a set of parallel, equi-spaced planes, so that the interval between rulings is not constant on the surface of the sphere, and the device, although it is a spherical mirror, has a definite optic axis.

The focusing property is as follows:

The grating is ruled on a spherical mirror of radius R. There is a circle, the Rowland circle, touching the sphere at its vertex perpendicular to the rulings and of radius $R/2$. The elementary theory of the grating states that monochromatic light from any point on this circle is diffracted according to the grating equation, and is subsequently focused at another point on the Rowland circle.

The theorem is not exact. There is no point focus but an approximation to the proof can be given by elementary geometry. In Fig. 9.1 the grating is displaced by an infinitesimal amount along the optic axis and intersects the Rowland circle at points x, x'. Suppose the grating constant at x is a. A ray coming from the centre of the Rowland circle at angle α to the axis meets the grating edge normally at x and is diffracted according to $\sin r = \lambda/a$. A similar ray directed along the grating axis meets it normally at o and is diffracted according to $\sin r' = \lambda/a'$. Both diffracted rays intersect the Rowland circle at y. Then by geometry,

$$\sin r = \frac{H}{2R}, \quad \sin r' = \frac{H}{r_g}, \quad \frac{R_g}{2R} = \cos \alpha.$$

---

[1] H. A. Rowland, *Phil. Mag.*, **16** (1883), 197.

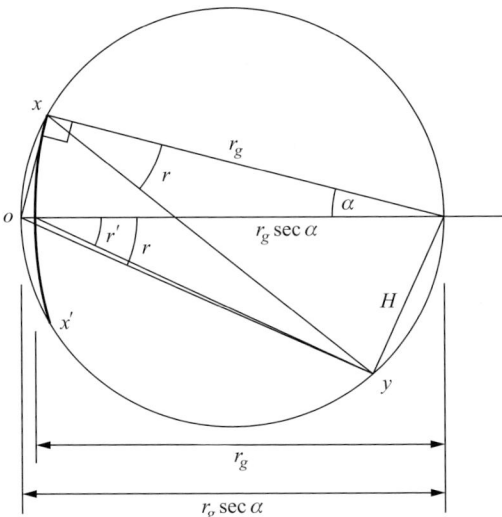

Figure 9.1 The basic geometry of the concave grating. The grating has been moved slightly (about 1 mm) to ease the geometry. Elementary diffraction theory predicts a focus at the point *y* provided the grating constant changes appropriately from the vertex to the edge.

By grating theory,

$$\sin r = \frac{\lambda}{a}, \quad \sin r' = \frac{\lambda}{a'},$$

and by elimination, $a = a' \sec \alpha$.

The focusing property therefore follows from the requirement for the spacing a to be constant along a chord. This argument, like the theorem, is not exact but the device works well enough within the limits of practical grating construction.

In practice it is the *tangential* focal surface which is supposed to lie on the Rowland circle, and a monochromatic point source is imaged as a line segment, long or short according to the dimensions of the grating.[2]

A theory of the focusing property of the Rowland grating is to be found, with some approximations, in R. W. Wood's *Physical Optics*.[3] However, the strict, three-dimensional theory is derived by Beutler[4] and a restricted form of this is presented here as an illustration of the different ways in which the grating may be used.

Beutler used the so-called 'eikonal' method in which the length of a ray is traced from an object point to an image point and measured in terms of the positions of

---

[2] To be more precise the tangential focal surface is a cylinder coaxial with the Rowland circle and containing it, and the sagittal focal surface is a similar cylinder turned through a right angle about the grating normal. These are in accordance with Eqs. (4.11) and (4.12).
[3] R. W. Wood, *Physical Optics* (New York: MacMillan, 1905).
[4] H. G. J. Beutler, *J. Opt. Soc. Am.*, **35** (1945), 311.

9.1 The Rowland grating

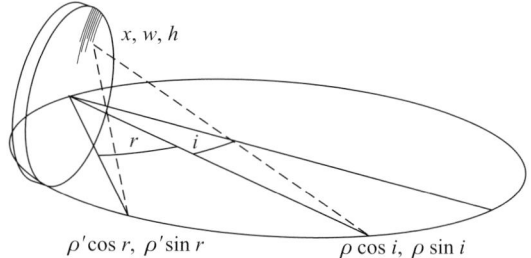

Figure 9.2 The geometry of the eikonal method used by Beutler for a strict description of the geometry of the concave grating. The method measures the path length from an arbitrary point on the Rowland circle to another arbitrary point where an image is formed, and the length is given as a function of wavelength, grating constant and the coordinates of the point of reflection. In the present analysis a simplified eikonal, ignoring the third dimension, is used.

the object and image and the coordinates of the intersection with the grating.[5] Polar and Cartesian coordinates are used as in Fig. 9.2.[6] The object and image positions are $(\rho, i, 0)$ and $(\rho', r, 0)$, the grating surface is $x^2 - 2Rx + y^2 + z^2 = 0$ and the intersection with the grating is (x, w, h). The condition for an image to be formed is that the optical path E shall change by one wavelength when the y-coordinate, w, of the intersection point changes by one ruling-width.

In the simplified eikonal used here, the z-coordinate h is neglected and the problem treated as two dimensional. The distance from object to image is then

$$E = \rho + \rho' - w(\sin i + \sin r) + \frac{m\lambda w}{a}$$
$$+ \frac{w^2}{2} \sum_0^\infty \left[\left(\frac{\sin i}{\rho}\right)^n \left(\frac{\cos^2 i}{\rho} - \frac{\cos i}{R}\right) + \left(\frac{\sin r}{\rho'}\right)^n \left(\frac{\cos^2 r}{\rho'} - \frac{\cos r}{R}\right) \right]. \quad (9.1)$$

The vanishing of various powers of w gives ever more stringent conditions for an image to be formed at the image point.

(1) The vanishing of the first power of w gives the standard grating equation $(\sin i + \sin r) - \frac{m\lambda}{a} = 0$.
(2) The vanishing of the second power gives

$$\left[\frac{\cos^2 i}{\rho} - \frac{\cos i}{R}\right] + \left[\frac{\cos^2 r}{\rho'} - \frac{\cos r}{R}\right] = 0. \quad (9.2)$$

---

[5] See e.g. M. Born & E. Wolf, *Principles of Optics* (Cambridge University Press, 1999), Chapter 3. Not always to be recommended because although it is precise, it involves miserably complicated and tedious elementary algebra and in practice yields little that is not available through Seidel theory.
[6] Beutler has the x-coordinate as the optic axis.

(3) The vanishing of the third power gives

$$\frac{\sin i}{\rho}\left[\frac{\cos^2 i}{\rho} - \frac{\cos i}{R}\right] + \frac{\sin r}{\rho'}\left[\frac{\cos^2 r}{\rho'} - \frac{\cos r}{R}\right] = 0, \tag{9.3}$$

and conditions (2) and (3) can be satisfied simultaneously by

$$\left[\frac{\cos^2 i}{\rho} - \frac{\cos i}{R}\right]\left[\frac{\sin i}{\rho} - \frac{\sin r}{\rho'}\right] = 0. \tag{9.4}$$

The first factor gives the Rowland condition. The second factor yields some alternative configurations which do not use the Rowland circle. This factor can be combined with Eq. (9.2) to yield

$$\rho = \frac{R[\cos^2 i \sin r + \cos^2 r \sin i]}{\sin r[\cos i + \cos r]} \tag{9.5}$$

and

$$\rho' = \frac{R[\cos^2 i \sin r + \cos^2 r \sin i]}{\sin i[\cos i + \cos r]}. \tag{9.6}$$

Two results follow:

(1) If $i = r$, then $\rho = \rho' = R\cos i$. This describes the *Eagle* configuration, where the image is formed above or below the entry slit.
(2) If $r = 0$, then $\rho = \infty$ and $\rho' = R/(1 + \cos i)$. This is the *Wadsworth stigmatic* configuration.

This latter is important because of its optical qualities. It is almost diffraction-limited and suffers only from spherical aberration and field curvature. A resolving power of 25 000 is possible with a grating of ruled area 60 mm × 80 mm. As with other arrangements of the Rowland grating it requires a curved focal surface, and this restricts it, for the moment at least, to silver halide emulsion photography.[7]

9.2 The concave grating as a spectrograph

9.2.1 The Rowland mounting

The original configuration was Rowland's and is shown in Fig. 9.3. He used the Euclidean theorem that the angle in a semicircle is a right angle, and mounted the grating cell on the end of a bar which could slide along a fixed rail. The other end of the bar was fixed to another rail at right angles to the first where the photographic plateholder was fixed. The entry slit was at the junction point of the two rails and

---

[7] There must in any case be some doubt whether a semiconductor detector could withstand the power of XUV radiation without irreversible changes to the crystalline structure of the detector.

9.2 The concave grating as a spectrograph

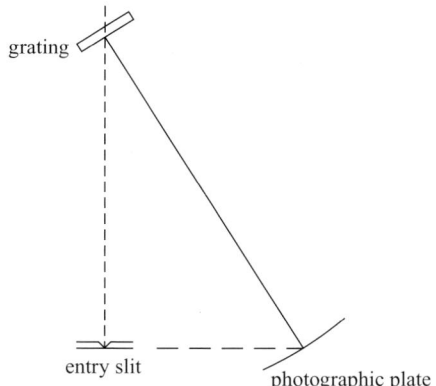

Figure 9.3 The Rowland mounting. The grating is fixed to one end of a bar of length equal to the grating radius. The photographic plate is fixed to the other end and the plate and the grating are face-to-face. The plate is then moved along a slide at right angles to another slide which constrains the grating to move towards the entry slit, thus preserving the illumination and étendue.

an optical bench for external optics ensured that the optic axis pointed directly at the grating vertex.

The distance between the grating vertex and the plateholder was equal to the diameter of the Rowland circle and the two pivots by which the bar was attached to the slippers were separated by the same distance. All this ensured that the grating was filled with light and that the photographic plate – which was bent on a mandrel in the plateholder to fit the Rowland circle – stayed in focus while the wavelength range was altered. This mounting still has its merits.

9.2.2 The Abney mounting

This is shown in Fig. 9.4. It was a variant on the Rowland mounting intended for use with very large spectrographs (and some concave grating spectrographs were on a huge scale, possibly filling a whole room) and very large plates. The plateholder and the grating were kept fixed and the input optics and source were moved from one entry slit to another when different regions of the spectrum were to be photographed. The inconvenience of this was so obvious that it was never particularly popular and is long consigned to history.

9.2.3 The Paschen–Runge mounting

The same is true of this mounting, also intended for large concave gratings: but now the input optics and grating are fixed on the Rowland circle and instead the

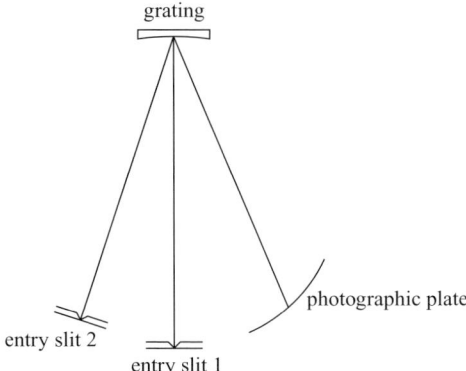

Figure 9.4 The Abney mounting. A relic of the time when spectrographs were sometimes constructed on – by modern standards – an enormous scale with focal lengths of 5 or 6 metres or more.

plateholder is moved on a railway track which runs around the circumference of the circle. This was another of the 'heroic' mountings mentioned in Chapter 1.

A smaller version of it with a 1-metre radius Rowland grating used a long curved plateholder with a strip of 35 mm film taped to a former of the same radius as the Rowland circle in a specially made film cassette. Some of these are still to be found in use.[8]

9.2.4 The Eagle mounting

In common with the Rowland mounting this uses the angle-in-a-semicircle theorem. In this case however, the plateholder is placed rigidly above (or below) the entry slit as in Fig. 9.5, and the bar serves to change the region to be photographed. The result is the most compact and useful configuration for the concave grating. It also serves as a monochromator mounting as described in the next section. The extra aberration resulting from out-of-plane working is negligible.

9.2.5 The Wadsworth stigmatic mounting

The Rowland focusing property is that the tangential focus lies on the Rowland circle. The sagittal focus at the same time is a plane tangent to the Rowland circle so that there is increasing astigmatism as the images are formed further and further from the grating normal.

---

[8] One at least, nearing 100 years old, was still in use at the end of the last century in the teaching laboratory in Manchester University.

9.2 The concave grating as a spectrograph

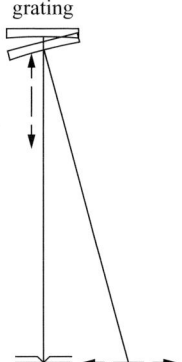

Figure 9.5 The Eagle mounting. This is a three-dimensional version of the Rowland mounting, and like it, uses the angle-in-a-semicircle theorem to maintain focus. The entry slit is above the plane of the diagram and the photographic plate, bent to fit the Rowland circle, is beneath it.

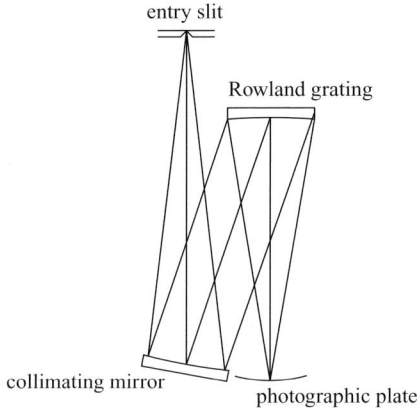

Figure 9.6 The Wadsworth stigmatic mounting. The Rowland focusing property is abandoned and a separate mirror is used as a collimator. The focal surface is a parabola and stigmatic images of the entry slit are formed on it at appropriate points. The main advantage, that the Rowland grating requires no other optical focusing element, is lost and this makes the mounting unsuitable for XUV spectrography.

The Wadsworth mounting[9] dispenses with the Rowland focusing property and illuminates the grating with collimated light (Fig. 9.6). If the grating is pivoted together with the plate or film holder attached to a bar on the grating normal, the line images are formed stigmatically on a paraboloidal focal surface to which the plate or film must be bent. The distance from the grating vertex to the focal

[9] F. L. O. Wadsworth, *Astrophys. J.*, **3** (1896), 54.

surface varies with the tilt angle away from the input chief-ray so that the position of the plateholder on the bar must be adjustable.

The mounting is awkward to adjust and focus and its advantage, that it is stigmatic, is that the images produced are more intense and that, as in old-fashioned spectrophotography, a Hartmann diaphragm can be used.[10]

There also remains the problem of the collimated input beam which in turn requires an extra optical element for collimation. This tends to remove the chief advantage of the Rowland grating: its freedom from other attenuators of XUV radiation.

9.3 The concave grating as a monochromator

This was the normal way of using the Rowland grating after the invention of the photomultiplier tube. XUV radiation, after diffraction, could be focused on to a fluorescent screen – usually a finely ground translucent layer of sodium salicylate – and the photocurrent from detection of the resulting visible radiation measured in the usual way, by a double-beam method if necessary.[11] Chiefly the Eagle arrangement was used but various Japanese authors published alternative linkages which relied on a single rotating vacuum joint instead of the cumbersome – and sometimes troublesome – linkage of the Eagle mounting.

First among these was the Seya–Namioka mounting[12] (Fig. 9.7), which held the slits fixed at two points on the Rowland circle separated by $\sim 141°$. Input and output chief-rays are thus separated by $70° 30'$. Tangential focus is then maintained as the grating is rotated about its own vertex. The Rowland focusing condition was clearly abandoned but there was the merit that good tangential focus could be maintained over a large range of wavelengths with no adjustment other than grating rotation. This property considerably eased the construction of vacuum spectrographs.

More practical was the Johnson mounting[13] modified by Onaka[14] in which the grating is rotated about an offset axis and the input and output chief-rays are separated by an angle 2ϕ. They describe an instrument with a grating of 40 cm radius, the angle 2ϕ as $30°$ and the offset of the rotation axis from the grating vertex as $l = R \sin\theta [1 - \tan\phi(\tan\alpha_0 - \tan\beta_0)/2]$.

---

[10] This is a device for wavelength calibration. It is a plate covering the input slit with three rectangular holes, staggered, in it. It can be slid across the slit so that the lower third, the middle third or the upper third of the slit can be photographed in succession, with a reference spectrum in the first and last cases. Line curvature sometimes makes it of dubious value.
[11] J. F. James, *J. Sci. Instrum.*, **36** (1959), 186.
[12] M. Seya, *Sci. Light*, **2** (1951), 8. Seya's mounting follows from the solution of a transcendental equation derived from the Beutler eikonal.
[13] P. D. Johnson, *Rev. Sci. Instrum.*, **28** (1957), 833.
[14] R. Onaka & I. Fujita, *Sci. Light*, **9** (1960), 31.

9.4 The aberrations of the Rowland grating

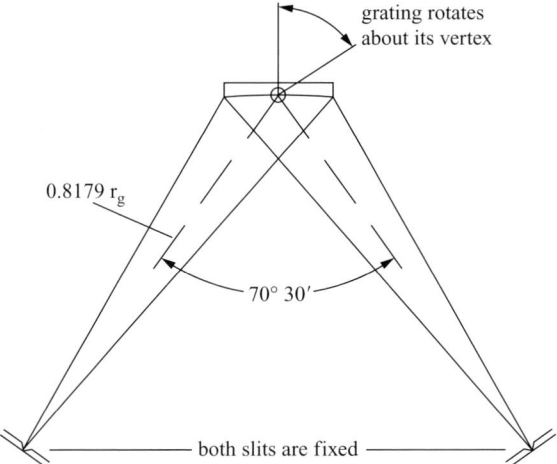

Figure 9.7 The Seya–Namioka mounting. This too abandons the Rowland focusing property, but, as discovered by Seya, two points are conjugate provided they are on the Rowland circle and separated by an angle of 70° 30′. Wavelength scanning is then by rotation of the grating about its vertex. The astigmatism however is a serious drawback with this type of mounting.

However, a great demerit appears when these designs are subjected to ray tracing. Given the usual length of grating rulings, there is a very large amount of astigmatism, up to half the length of the grating rulings, with its accompanying defocus. It is sufficient to dilute line images to an unacceptable degree and only a few millimetres of each ruling on either side of the plane of symmetry make any effective contribution to the measurable line images.

9.4 The aberrations of the Rowland grating

Astigmatism is the chief source of imperfection in any mounting. The full Beutler eikonal is required to discuss it, and if this is done an expression in closed form for astigmatism emerges. A point source on the Rowland circle with the usual angles of incidence and diffraction i and r, will produce an image at the tangential focus, also on the Rowland circle, of length $\mathcal{L}$ where

$$\mathcal{L} = l \left[\sin^2 r + \sin^2 i \frac{\cos r}{\cos i} \right] = \Lambda l, \tag{9.7}$$

where l is the length of the rulings. The coefficient Λ is required later.

9.4.1 Astigmatic curvature

The astigmatic image of a point source is not straight and neither is the image of a long input slit, for the reason outlined in Subsection 8.6.1. The two curvatures are in opposite directions and the overlapping images from different parts of the input slit yield a coma-like aberration which limits the resolution.[15] There are two cases to consider.

(1) Grating rulings of finite length and a point source on the Rowland circle.
(2) A long straight slit and infinitesimal grating rulings.

To simplify the analysis we define v by

$$\frac{1}{v^2} = [i \sin^2 i + \sin r \tan^2 r + \sin i \sin^2 r \tan^2 r - \sin 2i \sin r \tan r].$$

Then dr, the width of the line image, is in each instance:

(1) Long rulings, short input slit:

$$\mathrm{d}r = \frac{l^2}{2Rv^2} \quad \text{whence} \quad l = v\sqrt{2R\,\mathrm{d}r},$$

and l is the length of the longest rulings that can be used.

(2) Short rulings, long input slit:

$$\mathrm{d}r = \frac{s^2}{2Rl^2v^2} \quad \text{whence} \quad s = \Lambda v\sqrt{2R\,\mathrm{d}r},$$

where Λ is defined in Eq. (9.7) and s is the length of the longest slit that can be used.

It must be emphasised that these computations are based solely on geometry, and the effective resolution, dλ, in a concave grating spectrograph is obtained in each case from the usual formula $\mathrm{d}\lambda = (a/m)\cos r\,\mathrm{d}r$.

9.5 Practical details of design

There are some peculiarities of vacuum grating spectrography which are worthy of particular mention, beyond what is discussed in Chapter 14.

There are various ways of constructing vacuum spectrographs and some provide more practical difficulties than others. It is a sine qua non in vacuum technology that you have as few vacuum seals as possible, particularly if they are to be broken frequently.

So far as possible, remote operation of any interior moving parts by electric motor or magnetic coupling is to be preferred to a rotating vacuum seal. Low-voltage operation is essential, especially if low-pressure gases are to be tolerated or even required perhaps, in absorption spectrography.

---

[15] Coma itself depends on the coefficient of wl^2 in the Beutler eikonal, and as l is ignored, it does not appear here.

9.5 Practical details of design

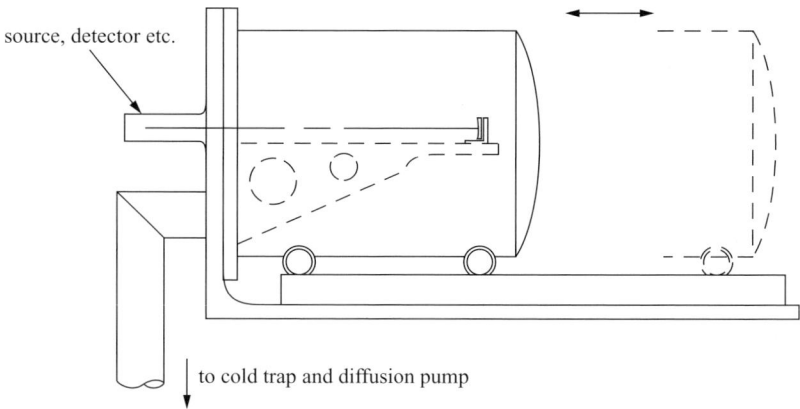

Figure 9.8 A possible arrangement for an Eagle mounting, with a separate vacuum chamber which can be removed for adjustments to be made to the optical system. All controls are led to the optical table through the end plate via vacuum seals which in general are not rotating and therefore leak-proof.

It is imperative that the instrument be designed so that it can be adjusted and focused in visible light. The awful alternative is to reduce the pressure to a vacuum each time a focus adjustment is made.

The author offers this advice from his own experience.

Build a monochromator rather than a spectrograph. Make it as small as possible, compatible with the research end in view. Choose the Eagle mounting, with a 1-metre radius grating with 6000 rulings/cm. Design for 10 mm-long slits centred 15 mm above and below the plane of symmetry. The defocus and coma are negligible out to 5000 Å. The resolving power is of the order of 5×10^4. Build the instrument as shown schematically in Fig. 9.8, on a horizontal table cantilevered out from a vertical face-plate which itself is carried on a suitable structure. Allow for a grating adjustment and traverse for a maximum wavelength of at least 5461 Å to permit visual adjustment of the optical alignment and focus. The table should have large holes for faster pumping. The vacuum chamber should be a simple cylinder able to run its whole length on horizontal rollers, wheels or linear bearings to make a single large-diameter vacuum joint at the face-plate. The seal may then be like that of a simple evaporator plant. This may be expensive, but saves much time in the long run.

Pass all controls through the face-plate in seals which need not normally be demountable. The face-plate also holds the slits, fore-optics and detectors. The lower half of the face-plate – below the optical table – holds the vacuum pump port, and the pump should have the largest pumping speed that the budget will bear.

It is unlikely that a vacuum lower than $\sim 10^{-6}$ mm Hg will be required, so that a turbo-molecular pump is probably unnecessary. Nevertheless for a 1-metre concave grating spectrograph in a 500-litre stainless steel chamber, a diffusion pump with at least 15–20 cm throat diameter should be included in the design. A cold-trap (LN_2) must be included.

10
The interference spectrograph

Interferometers were originally conceived as devices for measuring the wavelength of light or the precise measurement of small distances, and with spatial resolution and 'localised' fringes,[1] contour maps of reflecting surfaces could be made using Fizeau fringes or Tolansky fringes. They also served as very high resolution spectrographs with which the hyperfine structure of spectral lines could be studied. It was only later that their very high efficiency was noted and that they could be used with advantage at moderate resolution for the spectral examination of very feeble sources of radiation such as chemical phosphorescence, the night airglow and the zodiacal light.

10.1 The phase angle

There is an abstract angle, the 'phase angle', ϕ defined by

$$\phi = (2\pi/\lambda)2nd \cos\theta. \tag{10.1}$$

This is the most important general concept in interferometry, and describes all the varieties of interferometer. The phase angle comprises four variables: (1) the path difference, $2d$; (2) the angle of incidence, θ; (3) the incident wavelength, λ; and (4) the refractive index in the gap, n. The product $2nd$ represents the optical path difference between interfering wavefronts and $(2\pi/\lambda)2nd$ is the consequent phase difference.

Holding two of these four quantities constant and using a third as independent variable, the behaviour of the fourth describes one of a large number of possible fringe patterns. These include Fizeau fringes, constant inclination fringes, Mach–Zehnder fringes, Edser–Butler fringes, Fabry–Perot fringes, Tolansky fringes and

---

[1] There is a myth that in 'localised' fringes the interference takes place in the gap between two reflecting surfaces rather than 'at infinity' as in fringes of constant inclination. The interference of course takes place everywhere in space and in particular at the detector, in the retina of the eye or on the plate of the camera.

so on. Six types of interferometer result. So far as spectroscopy is concerned, λ is the dependent variable and can be made to depend on the variation of θ, d or n.

Interferometers for spectroscopy can be divided into two categories: spectrometers and spectrographs. In the former category there is a further subdivision into Michelson Fourier-transform spectrometers and étalon scanning monochromators, and in the latter category the Fabry–Perot spectrograph is the sole survivor of a century of experiment. Other devices such as Michelson's echelon and Lummer's plate, although they yielded very high resolution, were difficult to make, inconvenient to use and had far lower étendue than the étalon spectrograph. They are all now museum pieces.

10.2 The Fabry–Perot étalon spectrograph

The Fabry–Perot étalon was first described by Charles Fabry and Alfred Perot in 1899.[2] It comprises two flat transparent plates worked optically so that the gap between them can be held constant over the whole surface to an accuracy of better than $\sim\lambda/120$. The constant gap is the important aspect: the surfaces of the individual plates need not be flat to better than $\sim\lambda/4$.[3] When the device is dismountable, the angular orientation of the two plates may be important and is then indicated by a pair of arrows inscribed on the sides to ensure correct alignment. Etalons are often manufactured and assembled as sealed unitary devices with the gap permanently fixed so that no adjustment is required or possible, and these, dedicated to a particular investigation, are to be preferred when constructing a Fabry–Perot spectrograph.

The obvious advantage of a dismountable étalon is that the gap can be changed for different purposes and when this is needed it is desirable that there be a servo-stabilising mechanism for holding the gap parallel and constant during observations.

10.3 Fabry–Perot theory

The theory of the Fabry–Perot étalon is to be found in every textbook on physical optics. The transmitted intensity is given by the Airy formula:

$$I(\phi) = \frac{I(0)}{1 + \frac{4r}{(1-r)^2} \sin^2\left(\frac{\phi}{2}\right)}, \qquad (10.2)$$

where r is the reflection coefficient for intensity at the reflecting surfaces and, as before, $\phi = (2\pi/\lambda)2nd \cos\theta$. For values of δ close to zero, $\sin(\delta/2) \sim (\delta/2)$ and

---

[2] C. Fabry & A. Perot, *Ann. Chim. Phys.*, **16** (1899), 115.
[3] Although the modern tendency is to make the plates individually so that they can be assembled without regard to orientation.

10.3 Fabry–Perot theory

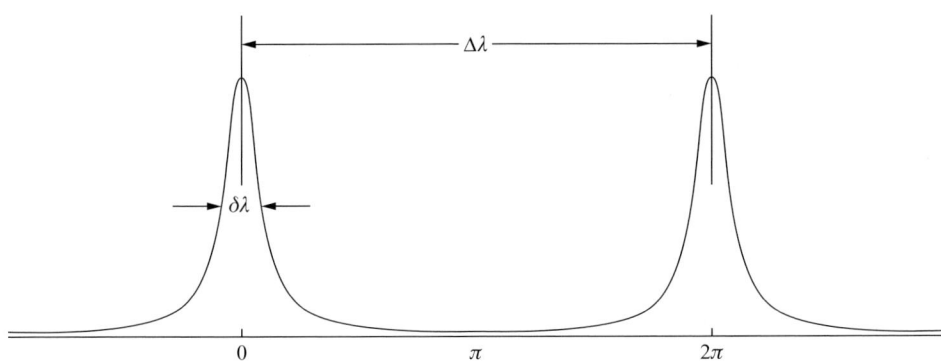

Figure 10.1 The Airy function of Eq. (10.2), illustrating the free spectral range $\Delta\lambda$ and the resolution, $\delta\lambda$.

the Airy function can be approximated by

$$A(\delta) = I(0) \frac{U^2}{U^2 + \theta^2},$$

where $U^2 = (1-r)^2/r$.

This is the equation of a Lorentz curve, and the Airy function (Fig. 10.1), which is periodic, may be regarded as the convolution of a Lorentz curve with a Dirac comb of period 2π. As we shall see in a moment, this Lorentz curve is the instrumental profile of the Fabry–Perot spectrometer and the number of such profiles that can be fitted into one order of interference is the ratio of the FWHM to the order spacing. The FWHM, expressed as an angle, is $2(1-r)/\sqrt{r}$ and the ratio of this to the period, 2π, is

$$F = 2\pi \frac{\sqrt{r}}{2(1-r)} = \frac{\pi\sqrt{r}}{1-r} \sim \frac{3}{1-r}.$$

This quantity F is called the *finesse* of the interferometer and is a measure of the number of resolved spectral elements in one order, that is, in one free spectral range (FSR). The finesse is affected by the quality of the reflecting coatings, by the wavelength to be examined and by the flatness of the gap. In addition there is a small effect caused by the polarisation of the light, as the effective gap width for oblique incidence – at the edge of the fringe pattern that is – depends on the state of polarisation, that is, whether the electric vector is radial or tangential. The effect, which vanishes on the optic axis, in the end controls the outer ring diameters and limits the available finesse.

For purely practical reasons, larger étalons of 50 mm diameter and above are likely to have a lower finesse than smaller ones of less than 25 mm aperture. Typical values in practice for a 50 mm étalon are $F = 35$–45.

The *free spectral range* $\Delta\lambda$ is the range of wavelengths that can be fitted between two orders of the same wavelength. The gap width, $d = n\lambda/2$, is also equal to $(n-1)(\lambda + \Delta\lambda)/2$, so that $\Delta\lambda = \lambda/n = \lambda^2/2d$.

The *resolution*, the smallest range of wavelengths, $\delta\lambda$, that can be observed as two distinct lines, is equal to the FWHM of the Airy function maximum, that is to $\Delta\lambda/F$, and can be written in terms of wavelength as

$$\delta\lambda = \frac{\lambda}{nF} = \frac{\lambda^2}{2dF} = \frac{\lambda^2(1-r)}{6d}$$

and r, the reflection coefficient of the plates, is usually in the region of ~ 0.93.[4]

10.4 The Fabry–Perot monochromator

10.4.1 Scanning in wavelength

This is done by changing the optical path between the two plates.

If the étalon is followed by a focusing lens the Fabry–Perot fringes from monochromatic light appear at the focus as fine concentric circles. An annular hole might be made in the screen with its inner and outer edges at the half-power points of a ring, and its diameters come from the Airy formula (Eq. 10.1) with the usual approximation $\cos\theta = 1 - \theta^2/2$. At the maximum of intensity,

$$\frac{\theta^2}{2} = 1 - \frac{n\lambda}{2d},$$

and at the half-power points,

$$\frac{\theta_1^2}{2} = 1 - \frac{n(\lambda - \delta\lambda/2)}{2d}, \qquad \frac{\theta_2^2}{2} = 1 - \frac{n(\lambda + \delta\lambda/2)}{2d},$$

whence

$$\frac{\theta_1^2 - \theta_2^2}{2} = \frac{n\delta\lambda}{2d}$$

$$= \left(\frac{n\lambda}{2d}\right)\left(\frac{\delta\lambda}{\lambda}\right) = \frac{\cos\theta}{R}.$$

It is usual – but not universal – to let the inner diameter of the annulus go to zero so that the wavelength-defining stop is a circle. Then $\theta_2 = 0$ and $\cos\theta_1 \simeq 1$ and the angular *diameter*, θ, of the circular stop is given by

$$\frac{\theta^2}{8} = \frac{1}{R}. \tag{10.3}$$

---

[4] With multilayer reflecting coatings this high reflection coefficient holds good only over a hundred angstroms or so, and must be specified when ordering the plates.

10.4 The Fabry–Perot monochromator

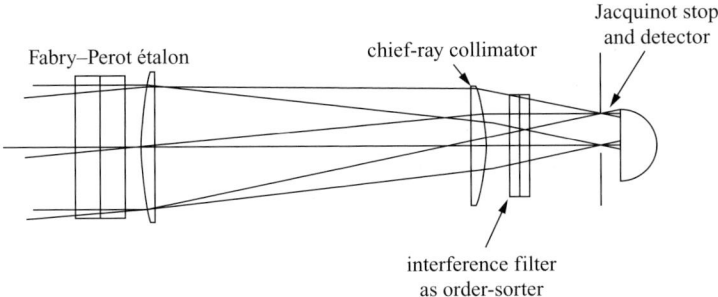

Figure 10.2 The classical mounting for a Fabry–Perot spectrometer. The incoming light, which must be from an extended source, is focused on to the Jacquinot stop but first passes through a short section where the chief-rays are collimated before passing through the interference filter which is the order-sorter in this type of arrangement. Wavelength scanning is by one of various methods of changing the optical path between the plates. Fore-optics, usually in the form of an astronomical telescope, could be added if the source was of small extent or if fine detail in the source was to be examined.

These relationships between the various quantities λ, d, r and θ can be used to define the important quantities associated with the Fabry–Perot spectrometer and spectrograph.

(a) The free spectral range (FSR) = $\Delta\lambda$, $= \lambda/n$, $= 2d/n^2$, $= \lambda^2/2d$.
(b) The finesse, $F = \Delta\lambda/\delta\lambda$, $= \pi\sqrt{r}/(1-r)$, $\sim 3/(1-r)$.
(c) The resolving power $R = \lambda/\delta\lambda$, $= nF$, $= 2Fd/\lambda$.

A typical arrangement for a scanning Fabry–Perot spectrometer is shown in Fig. 10.2. There is a short section near the Jacquinot stop where the chief-rays are all made parallel to the optic axis and this is where the order-sorting interference filter is placed. The filter, which is a Fabry–Perot étalon working at a very high finesse and a very low order, requires a divergence less than F/8 or higher (the higher the better), and this focal ratio should be built-in as part of the initial design.

10.4.2 Mechanical scanning

The mode of mechanical displacement of one plate is generally by piezo-electric transducers, mounted typically as in Fig. 10.3. An individual piezo-electric[5] ceramic disc of barium titanate, for example, will change its thickness by ~ 3 Å V^{-1} applied across its faces and a stack of ten such ceramic discs of alternating polarity will multiply this by 10, to 30 Å V^{-1}. A scanning range of one order is thus achieved with a ramp of ~ 100 V, a ramp function which can be repeated slowly over 1500 s, or rapidly at 50 Hz or more. The efficiency and the signal/noise ratio obtained is the

---

[5] *Not* electrostrictive, but a disc which has been electrically polarised during manufacture.

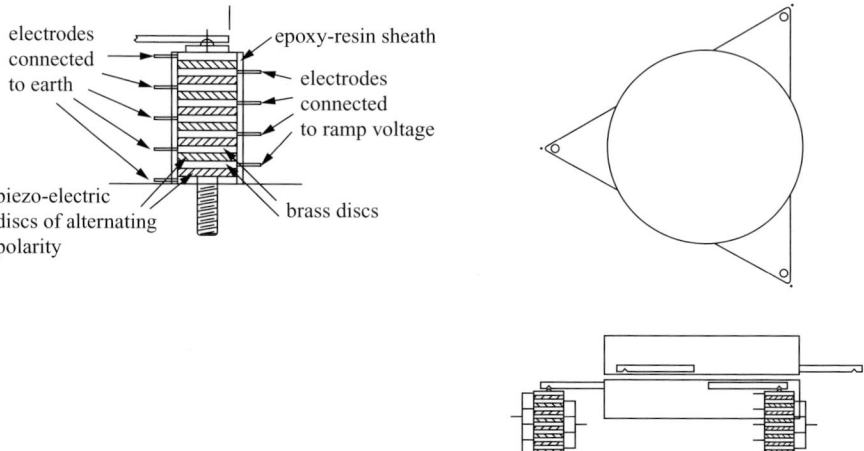

Figure 10.3 The mounting of piezo-electric feet for the plate when mechanical scanning is used to change the gap. Control of the gap width was either by capacitance measurement or by transmission of Edser–Butler fringes through opposite edges of the gap.

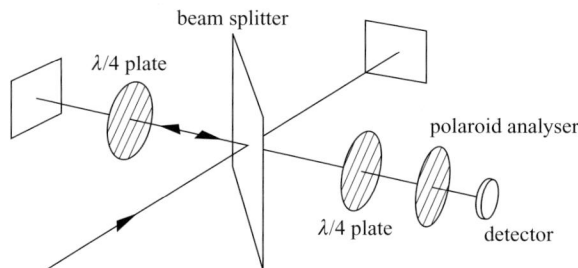

Figure 10.4 An optical screw for controlling and changing the gap. This device is also applied to the gap in Michelson Fourier spectrometers and allows changes of the thickness of $\lambda/500$ to be made under computer control.

same in both cases and both scan rates may have advantages. The voltage applied can be controlled by an optical screw which changes the gap by one order (i.e. by $\lambda/2$) for every half-turn of the screw.[6]

The mechanism is simple: a stabilised laser (Fig. 10.4) sends a plane-polarised beam into a Michelson interferometer with the plane of polarisation parallel to the beam-splitter surface. One beam is reflected and returned to the output port while the other passes through a quarter-wave plate and is converted to circularly polarised light. After reflection it passes again through the quarter-wave plate and is then converted back to plane-polarised light but now with the plane of polarisation perpendicular to the original plane. It too passes through the output port, but although

---

[6] Y. P. Elsworth & J. F. James, *J. Phys. E: Sci. Instrum.*, **6** (1973), 1134.

it is coherent with the other output beam there is no interference because of the mutually perpendicular planes of polarisation. The two beams now pass together through another quarter-wave plate, and one is converted to left circularly and the other to right circularly polarised light. The result is plane-polarised light, but the plane of polarisation depends on the path difference in the interferometer. A piezoelectric transducer vibrating on one of the Michelson reflectors causes a minuscule sinusoidal variation in the path difference, enough to cause the plane of polarisation to oscillate in direction and a phase-sensitive detector measures the amplitude and causes a servo-mechanism to turn an analyser to the extinction direction. The angle of the analyser indicates the path difference. Alternatively the analyser can be turned and the path difference changed to match by a servo-mechanism in the interferometer. The interferometer is very small – large enough to allow a laser beam to traverse – and one of its reflectors is a small patch on the étalon plate.

During scanning the flatness of the gap can be controlled by two types of servo-mechanism. In one case[7] a narrow collimated beam of white light is directed through the edge of the étalon and Edser–Butler fringes would be observed if the light were analysed. Instead it is passed a second time through the opposite edge and there will be full transmission of the Edser–Butler fringes only if the two gaps are exactly equal. If one plate is made to rock through a small amplitude – a few angstroms is sufficient – the transmitted intensity varies rapidly and again a phase-sensitive detector tuned to the rocking frequency will sense the direction and a voltage amplifier will apply a restoring force to the transducer. Two such servo-loops, mounted normally to each other, will stabilise the gap in two dimensions.

Alternatively there is a mechanism described by Hicks et al.[8] Two small capacitors can be made by aluminising patches at opposite points on the étalon rim, and the two capacitances ($\epsilon_0 A/d$), will be the same only if the gap d is constant. An a/c bridge then measures any change in the two capacitances and applies an appropriate restoring movement.

10.4.3 Pressure scanning

This, like mechanical scanning, describes a scanning monochromator, but now the étalon is held in a pressure chamber. The gap width is held constant and the same diameter of circular stop defines the wavelength range transmitted. The refractive index, n, of the gas in the chamber is changed slowly by a pressure valve according to $n = (1 + kp)$. For air k typically is 0.00011 atmosphere^{-1} but may be higher for various halogen-substituted hydrocarbon vapours (CFCs). The method is thoroughly redundant and obsolete and not to be considered further.

---

[7] J. V. Ramsay, *Appl. Opt.*, **1** (1963), 411; M. J. Smeethe & J. F. James, *J. Phys. E: Sci. Instrum.*, **4** (1971), 429.
[8] T. R. Hicks, N. K. Reay & R. J. Scaddan, *J. Phys., E*, **7** (1974), 27.

10.4.4 Multislit scanning

The efficiency of the scanning monochromator can be improved by an order of magnitude by using a focal plane mask with concentric annular apertures.[9] The instrument must be designed and built before the mask can be made, since the angular diameters of the apertures must coincide with the ring system. The mask is conveniently made by setting the instrument to look at a monochromatic source with a wavelength close to the wavelength to be observed, then to photograph the Fabry–Perot fringes with a high-contrast, fine-grain emulsion on a glass plate. The mask can then be made by 'printing' the negative or by reversal development, in which the original reduced silver image is bleached away and the remaining emulsion exposed and developed. Up to ten rings can be included in the mask before the polarisation effect mentioned above interferes seriously with the finesse.

10.5 The Fabry–Perot CCD spectrograph

The original Fabry–Perot spectrograph was abandoned for serious research when the photomultiplier arrived in the 1950s. Since the CCD reversed the situation the Fabry–Perot spectrograph now leads the field in this kind of interference spectroscopy. The advantages are manifold. The étalon is fixed, and such a sealed, unitary device is ideal for its purpose where the resolving power required is known beforehand. There are then no moving parts – other than electrons – to the spectrograph. Although photography may still be by silver halide emulsion, the CCD is the detector of choice whenever fast transient phenomena or very feeble light sources are to be examined.

The efficiency of the device is determined by the area of the CCD multiplied by the solid angle subtended at the étalon aperture, or equivalently, by the square of the numerical aperture of the camera lens. The statement is subject to various constraints. These are:

(1) There must be at least one complete ring diameter photographed, and in practice 2 or $2\frac{1}{2}$ rings are desirable. The rings need not be complete, because this depends on the aspect ratio of the CCD chip, but a means of determining the x- and y-coordinates of the pixel at the centre of the ring system must be available.

The ring diameters, D_n, are given by the well-known formula

$$D_n^2 = 8f^2(1 - n\lambda/2d), \qquad (10.4)$$

where n refers to the nth ring from the centre of the field, d is the gap width in the étalon and f is the camera lens focal length. At the centre of the ring system, n is given by

---

[9] Amal Al-Hillou, Ph.D. Thesis, University of Manchester (1976).

$n = 2d/\lambda$ and is not usually integer. Remember that the order, n, decreases as the ring diameters increase.

(2) The outer ring thickness for strictly monochromatic light must be several (at least 3 or 4) pixels, otherwise the spectral resolution suffers.
(3) The camera lens must have an aperture equal to the étalon clear diameter and the highest numerical aperture available.
(4) There must be provision for an order-isolating filter. This is usually an interference filter with a FWHM bandwidth equal to half the free spectral range of the étalon. Such a filter requires that all chief-rays passing through it must be parallel to the optic axis. This is to ensure that the same spectral range is photographed at all points of the detector, remembering that the filter itself is a Fabry–Perot étalon with an effective gap $\sim 1/40$ that of the principal étalon.

The precursors of the design are:

(1) The resolution $\delta\lambda$ required. This determines the gap thickness, which in turn determines the angular diameter of the first two rings.
(2) The dimensions of the CCD chip. This determines the focal length of the camera lens, which must be such as to allow at least two and not more than three complete rings to fall on the CCD.
(3) The numerical aperture of the necessary lens. This determines the diameter of étalon required *after the optical paths have been laid out*. The diaphragm of the camera lens and not the étalon is the pupil of the system and enough clearance between the front surface of the lens and the rear surface of the étalon must be given.

10.5.1 The L-R product

Like the scanning Fabry–Perot spectrometer, the spectrograph has an L-R product and it is greater than that of the scanning spectrometer by a factor $16F$, where F is the finesse of the étalon.

This can be derived from the need for two rings on the CCD. If the order of interference is n_0 at the centre of the ring system it must be $(n_0 - 2)$ at the edge of the CCD chip, and the diameter of this ring is connected to the resolving power by

$$D_2^2 = 8f^2 \left[1 - \frac{(n_0 - 2)}{n_0}\right] = \frac{16f^2}{n_0} = \frac{16f^2 F}{R}.$$

D is the width of the CCD chip and its area is D^2 if it is square (or proportional to D^2 if it is not), so that with an étalon of diameter D_e, the étendue of the whole spectrograph is

$$L = \frac{\pi D_e^2 D^2}{f^2} = \frac{\pi D_e^2}{f^2} \frac{16f^2 F}{R},$$

and the efficiency, $E = LR = 16\pi D_e^2 F$.

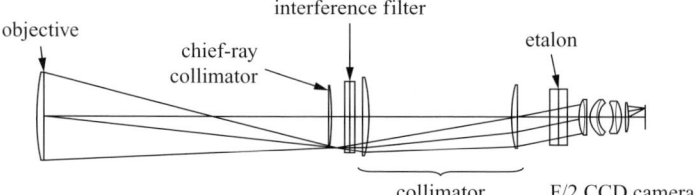

Figure 10.5 The Fabry–Perot CCD spectrograph. A telescope appropriate to the extent of the object under examination produces an image of the object at its focus and, as before, a short section is included where the chief-rays are collimated before passing the order-sorting interference filter. The object is then re-collimated before passing through the étalon. The auxiliary function of the optical system is to form an image of the objective on the camera pupil and an image of the object on the CCD. The evolution of this type of spectrograph is discussed in more detail in Appendix 3.

10.6 Fore-optics

Figure 10.5 shows a Fabry–Perot spectrograph designed for airglow spectrography. The telescope objective is followed by a chief-ray collimator near its focus. This is a lens of the same focal length as the objective and it serves to pass all ray bundles through the interference filter in the same orientation. The light is then re-collimated preparatory to its passage through the étalon and simultaneously the objective is imaged on to the CCD camera pupil. Two lenses are employed to avoid using an inconveniently large collimator. The evolution of this design is described as a 'worked example' in Appendix 3.

10.7 Reference fringes

It is generally a good idea to record a reference wavelength at the same time as the spectrum under investigation. One method is to use a 'keyhole' projector, imaging the refernece source through a spatial filter on to a small Newtonian mirror on the optic axis of the spectrograph (Fig. 10.6). The reference wavelength, from a ^{198}Hg lamp for instance, need not be within the pass-band of the order-sorting interference filter but it must be within the wavelength range for which the étalon coatings were made. To avoid confusing the reference fringes with details in the observed spectrum, the light from the reference source may be passed through a spatial filter to interrupt the fringes it produces. The light coming through the filter is collimated and an image appears on the CCD. Since no sharp image of the mask is needed, the projector optics may be simple plano-convex lenses.

10.8 Extraction of the spectrum

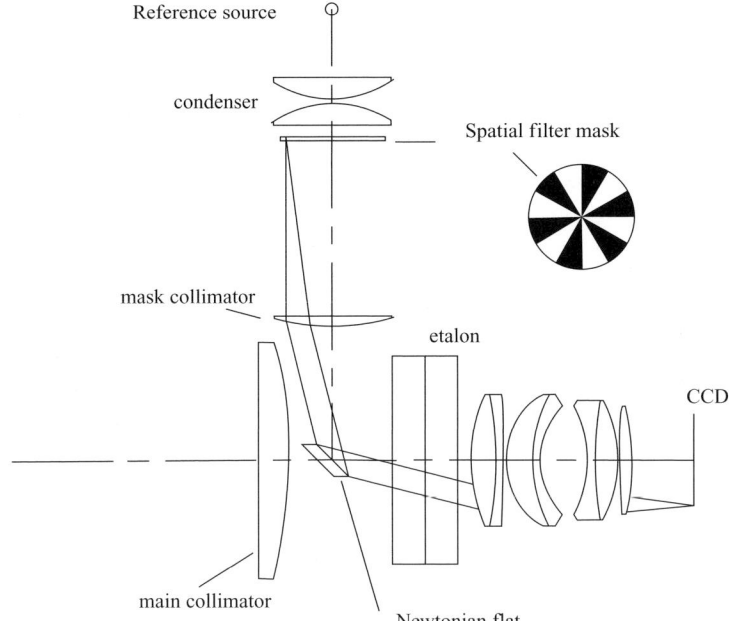

Figure 10.6 The reference fringe injector uses 'keyhole' optics. The condenser images the reference source on to the Newtonian mirror. The mask is at the back focus of its collimator, so that at the camera it appears to be at $-\infty$ and is imaged with interrupted reference fringes at the detector. No elaborate optical system is required and only a few per cent of the étalon area need be obscured by the Newtonian mirror.

10.8 Extraction of the spectrum

The CCD output will be in the form of a two-dimensional array of numbers representing the content of each pixel at read-out time. At the time of writing the FITS format is popular and adequate because the contents of each pixel can be read and, if necessary, altered individually.[10]

Finding the pixel coordinates of the centre of the ring system can be done by software, but in practice it is a simple task to do so adequately by inspection.

The following software procedure will extract the spectrum:

(1) Three linear arrays are required in the computer program.
(2) The two-dimensional read-out of the field is inspected visually and the pixel at the centre of the ring system is identified. A computer program can be written to do this, but in practice a simple visual inspection will find the appropriate pixel.

---

[10] **F**lexible **I**mage **T**ransport **S**ystem: visit an internet site for details of its format.

(3) Each pixel is taken in turn and the square of its distance in pixels from the central pixel is computed and used as the address in all three arrays. In general this number will not be an integer and the array should be long enough (i.e. should have enough addresses) that the integer part of the calculated number or its multiple can be used. (For example, if d^2 comes to 102.776, this can be made to correspond to address number 1028.)
(4) The contents of the pixel are stored at the corresponding address in array [a].
(5) The number in the corresponding address in array [b] is increased by 1 to indicate the number of times the intensity at that distance from the centre has been added to array [a]. This is to make allowance for incomplete rings that may be on the CCD image.
(6) Array [c] is constructed containing array [a] divided by array [b] and this contains the average intensity at (the square of) each distance from the centre. The addresses are numbers proportional to the wavelength at each ring via Eq. (10.3) above. A graphical plot of array [c] is then a linear plot of the required spectrum, repeated one order later. The geometry of the spectrograph allows the wavelength corresponding to each address to be computed.

The method is particularly designed to cope with incomplete rings, but it happens from time to time that there is need for spatial resolution in the interferogram. The corona of the Sun or a luminous interstellar cloud may be in focus on the CCD together with the rings. In this case the same reduction technique may be used, but with the area restricted to correspond to a feature or a particular small area in the source.

10.9 Choice of the resolution and gap

This depends on the nature of the spectrum to be investigated. For hyperfine structure, then clearly the resolution must be sufficient to resolve the various components. For small changes such as Doppler shifts, the intention must be to maximise the change that appears and this in turn depends on the spatial frequencies present in the displayed spectrum. Higher resolution implies less luminosity and hence a longer observing time, and this, with faint sources, could be the limiting factor. There is no point in using more resolution than is needed. The 'matched filter theorem'[11] guides the gap selection here. The Fourier transform of the instrumental profile of a Fabry–Perot spectrograph is a Dirac comb multiplying a decaying exponential:

$$\mathrm{III}(p)\mathrm{e}^{-p/b} \rightleftharpoons \mathrm{III}(x) * \frac{1}{4\pi^2 x^2 b^b + 1},$$

the right hand side being the instrumental profile of the étalon. If the observed spectrum line is regarded as a Gaussian of FWHM w then its shape is represented

---

[11] 'The signal/noise ratio in a receiver is greatest when the noise-filter transmission profile has the same shape as the power spectrum of the incoming signal.' Although this was derived originally for radio transmission, it is quite general and applies throughout information and communication theory, including spatial filtering.

by
$$e^{-1.92x^2/w^2} \rightleftharpoons e^{-\pi^2 p^2 w^2/1.92}.$$

These two curves can not be overlaid and matched exactly but a reasonable approximation is that they cross at the 1/e-point. Then,

$$\pi w/1.386 = 1/b$$

and the FWHM of the Fabry–Perot instrumental profile should be that of the observed line FWHM divided by 1.386.

This is still approximate, and in practice much the same efficiency is obtained provided that the étalon line width is about the same as the observed line or somewhat less (rather than more). It is impossible to be more precise as the line shapes being observed are not always conveniently Gaussian and the spectrum may contain lines of different half-widths.

10.10 The 'crossed' Fabry–Perot spectrograph

A technique for improving the resolution of emission line spectra in a spectrograph by an order of magnitude was developed in the early days by imaging the rings of a Fabry–Perot étalon on to the entry slit of a Littrow spectrograph, the centre of the ring system lying on the line of the slit. The slit was then opened wide, subject to not allowing line images at different wavelengths to overlap and the spectrogram revealed such things as pressure broadening, Zeeman splitting and hyperfine structure simultaneously at a large number of wavelengths – for example in a molecular band spectrum. The technique depends essentially on having small or zero astigmatism in the grating or prism spectrograph and is therefore most suitable for the Littrow configuration. The image of the ring system should contain both parts of the innermost ring in order that the centre of the system be located.[12] The same criteria apply as for a normal Fabry–Perot spectrograph: the number of rings should be at least two, and with a CCD detector the instrumental profile should be at least 5 pixels wide at FWHM if the line profile is required.

The technique is not yet moribund and, if applied, the same rules apply for imaging the source with the proviso that if the source has structure of its own – an oxy-acetylene flame for example – the magnification and defocus should be such that the local variations do not affect the interferogram.

---

[12] The alternative is a tedious amount of programming to linearise the square-law wavelength dispersion along the slit direction.

11
The multiplex spectrometer

The theory and practice of Michelson Fourier-transform spectrometry is strictly speaking beyond the purview of this book, but it is in such universal use that a brief description is warranted in order that a reasoned choice can be made when contemplating infra-red spectroscopy.

11.1 The principles of Fourier spectrometry

When a monochromatic beam of light passes through an ideal Michelson interferometer with no absorption and exactly 50% reflection at the beam-splitter, the intensity of the beam which emerges depends upon the path difference between the two recombined components.

Suppose that the incident beam has wavenumber ν and complex amplitude A. Suppose, for simplicity, that the phase is zero at the beam-splitter. After passing the beam-splitter the two emerging amplitudes will be

$$\frac{A}{\sqrt{2}}e^{i\delta_1} \quad \text{and} \quad \frac{A}{\sqrt{2}}e^{i\delta_2},$$

where δ_1 and δ_2 are the phase changes on reflection and transmission.

After passing through the two arms of the interferometer the two beams recombine at the beam-splitter. Before recombination the two amplitudes are[1]

$$\frac{A}{\sqrt{2}}e^{i\delta_1}e^{2\pi i\nu d_1} \quad \text{and} \quad \frac{A}{\sqrt{2}}e^{i\delta_2}e^{2\pi i\nu d_2},$$

and after recombination, when each beam has undergone one transmission and one reflection, the resultant amplitude transmitted is

$$\frac{A}{2}e^{i(\delta_1+\delta_2)}\left(e^{2\pi i\nu d_1} + e^{2\pi i\nu d_2}\right).$$

---

[1] Bearing in mind the precept that phase change = $2\pi\nu \times$ path change.

11.1 The principles of Fourier spectrometry

This quantity, multiplied by its complex conjugate gives an expression for the transmitted intensity:

$$I(\Delta) = \frac{AA^*}{2}[1 + \cos(2\pi\nu\Delta)],$$

where $\Delta = 2(d_1 - d_2)$ is the path difference between the two beams.

The transmitted intensity can be seen to vary sinusoidally with Δ. The intensity and the frequency of this wave depend on the rate at which the path difference changes and on the intensity and wavenumber of the incident radiation.

In practice the input has a continuous spectrum $S(\nu)$ and if the intensity in the range $\nu \to \nu + \delta\nu$ is $S(\nu)\,d\nu$, it will contribute an infinitesimal power $dI(\Delta) = (S(\nu)\,d\nu/2)[1 + \cos(2\pi\nu\Delta)]$ to the output beam.

The total transmitted intensity will then be

$$I(\Delta) = \frac{1}{2}\int_{\nu=0}^{\infty} S(\nu)\,d\nu[1 + \cos 2\pi\nu\Delta]$$

$$= \frac{1}{2}\int_{\nu=0}^{\infty} S(\nu)\,d\nu + \frac{1}{2}\int_{\nu=0}^{\infty} S(\nu)\cos 2\pi\nu\Delta\,d\nu$$

and the second term is the Fourier cosine transform of the spectral power density, $S(\nu)$, as a function of Δ.

An alternative way of looking at Fourier spectrometry is to regard the signal strength at the detector as the time average of the product of the signal and its delayed counterpart:

$$I(t') = \langle [f(t) + f(t+t')][f^*(t) + f^*(t+t')] \rangle$$
$$= \langle f(t)f^*(t) \rangle + \langle f(t+t')f^*(t+t') \rangle + 2\langle |f(t)f(t+t')| \rangle,$$

where $t' = \Delta/c$ and Δ is the path difference. The third term is the autocorrelation function of $f(t)$ and by the Wiener–Khinchine theorem its Fourier transform is the SPD, or the spectrum of the signal.

The basis of Fourier-transform spectroscopy is that the output signal from the interferometer is recorded as a function of Δ (the *interferogram*) and the spectrum is obtained from this by a Fourier transform, usually in a digital computer.

The sampling theorem is used, and samples of the transmitted intensity are taken at equal intervals of path difference. The number of samples may vary from 1024 to 10^6, depending on the investigation being followed. This set of numbers takes the place of the interferogram and is sent as a one-dimensional array or *vector* to the computer.

The art of multiplex spectrometry lies in three directions.

Firstly, the optical alignment must be maintained to interferometric accuracy during the whole displacement phase of several centimetres or more while the interferogram is being recorded.

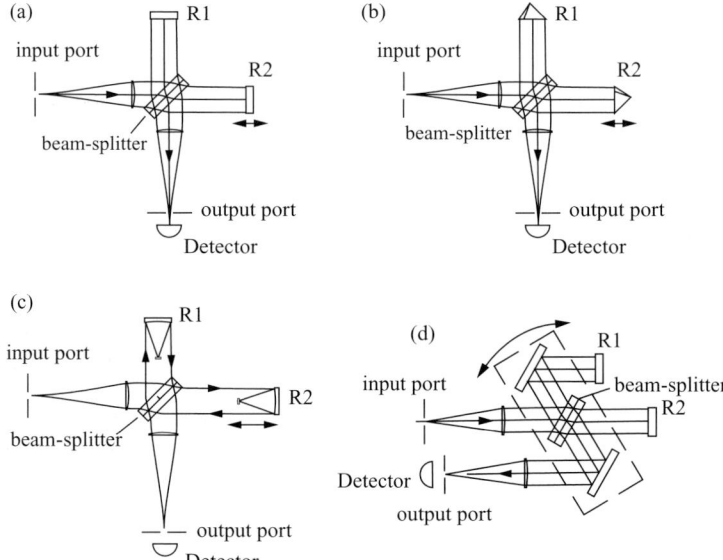

Figure 11.1 Variations on a theme. Four different types of Michelson Fourier spectrometer which have found commercial employment in different wavelength regions. (a) The classical Michelson interferometer; (b) the same, with cube-corner reflectors for automatic alignment maintenance; (c) the 'cat's-eye' autocollimator reflector which returns any incidence ray anti-parallel to its input direction; (d) the 'tilting' Michelson, in which the path difference is changed without physically moving the reflectors.

Secondly, the samples must be collected at exactly equal intervals of path difference. The process is analogous to ruling a new diffraction grating every time a spectrum is observed, and the precision required is the same as that of a ruling engine,[2] although the time devoted to each spectrum is considerably less.

Thirdly, an efficient way of doing the Fourier transform mechanically is required.

Much research was devoted to these three problems in the 1960s and 1970s and various solutions were found. In far IR spectroscopy, simple mechanical displacement of a reflector is adequate. At shorter wavelengths, cat's-eye reflectors and cube-corner reflectors allow linear translation of the Michelson mirrors without misalignment, and the 'tilting Michelson' changes the path difference by rotating a jig containing the beam-splitter and a rigorously parallel mirror about a hinge. Examples of successful applications of these ideas in manufactured Fourier spectrometers are shown in Fig. 11.1.

If samples of the interferogram must be taken when the reflectors are stationary, piezo-electric transducers can be made to oppose a smooth steady change in path

---

[2] The late Dr H. J. J. Braddick once remarked that designing a grating-ruling engine was the same problem as designing a screw-cutting lathe to be made of rubber.

difference with servo-stabilisation of both the alignment and path difference for each sample. Wavelength-stabilised lasers offer a standard of step length either by ensuring that the step length is a constant number of half-wavelength fringes or, by an optical screw[3] or ratchet device, a constant fraction (often vulgar) of a fringe.

The Fourier transformation of the interferogram is by digital computer, usually a dedicated laptop, using the fast Fourier transform or FFT, one of a number of routines derived from the Cooley–Tukey algorithm of 1964.[4]

The reason for doing spectroscopy in this way is the enormous gain in signal/noise ratio over grating spectroscopy which is obtained when the dominant noise source is the detector noise. This follows from P. J. Fellgett's discovery in the 1950s of the multiplex advantage.[5] This merits some discussion as it is typical of the way in which detection methods are assessed.

11.2 The multiplex advantage

Consider a grating spectrometer and a multiplex spectrometer of the same étendue E and observing the same spectral source $S(\nu)$, in each case to be explored over a bandwidth ν_0. The spectrum is everywhere dense and the total power is S.

Suppose each instrument dwells at each sample point for a time T. Suppose there are to be N resolved elements in the spectrum.

In the monochromator the flux to the detector is

$$F_g = \frac{E S \nu_0 T}{N},$$

which is the total signal energy in each sample. The detector noise is proportional to $\sqrt{T}$ and the signal/noise ratio achieved is

$$S/N = \frac{E S \nu_0 T}{N k \sqrt{T}}$$
$$= \frac{E S \nu_0 \sqrt{T}}{N k}.$$

In the multiplex spectrometer the total signal in each sample is much greater:

$$F_m = \frac{E S \nu_0 T}{2},$$

and again the noise energy at each sample is $k\sqrt{T}$.

---

[3] Y. P. Elsworth & J. F. James, *J. Phys. E: Sci. Instrum.*, **6** (1973), 1134.
[4] J. W. Cooley and J. W. Tukey, *Mathematics of Computation*, **19** (1965), 297. See also: E. Oran Brigham, *The Fast Fourier Transform* (Prentice-Hall, 1974).
[5] P. Fellgett, *J. de Phys.*, **19** (1958), 187, 237.

In this case the samples are used to compute the energy in each element of the spectrum by

$$S(m) = \sum_{n=0}^{N-1} \left(\frac{ESv_0T}{2}\right) \cos\frac{2\pi mn}{N},$$

and since cosines are as often negative as positive the total energy in the computed sample will be

$$F_m = \frac{ESv_0T}{2}.$$

The noise samples, also negative as often as positive will average to 0 but the RMS value will be

$$\sqrt{\sum_{n=0}^{N-1} k^2T \cos^2\frac{2\pi mn}{N}} = \sqrt{\frac{Nk^2T}{2}} \quad \text{since} \quad \langle\cos^2\theta\rangle = \frac{1}{2},$$

and the signal/noise ratio is then

$$S/N = \frac{ESv_0}{2k}\sqrt{\frac{T}{N}},$$

which is greater than that of the monochromator by a factor $\sqrt{N}/2$.

A similar calculation for the same spectrometers with photoelectron shot-noise as the principal source of noise shows that there is no such multiplex advantage and that the S/N ratio is apparently the same for each type of instrument. The calculation however does not show that there is in fact a multiplex *disadvantage* since the shot-noise on any one strong emission line is spread throughout the interferogram and hence throughout the spectrum, and it effectively drowns any faint emission lines which may be present. Similarly, the shot-noise from a continuous emission spectrum fills in or obscures any narrow shallow absorption lines which may be present.

The above calculations are not exact since source strengths may vary from point to point in the spectrum as well as from moment to moment during the observation, and the multiplex advantage is usually present only in order of magnitude. When the number of samples (or the resolving power) is high, $\sim$100 000 for example, the multiplex advantage is $\sim$300, the time taken to obtain the spectrum is reduced from 5 minutes to 1 second, and in the infra-red beyond 1 μm, where sensitive area detectors still await invention, the Michelson Fourier spectrometer is the instrument of choice.

Fourier spectroscopy is most suited to absorption spectroscopy where (usually) vapour samples are observed with a constant, well-regulated infra-red source

11.2 The multiplex advantage

providing the radiation power, in which case it has a marked advantage over grating spectroscopy both in accuracy and time. It is only to be undertaken on variable sources when the frequency spectrum[6] of the variation does not include the sampling frequency or, more prudently, when a separate detector monitors the moment-to-moment intensity fluctuations of the source. But beware: this latter precaution does not work if the spectral content varies from moment to moment. It is the sort of thing which happens in atmospheric research for example, by the selective absorption of sunlight in the water vapour of occasional passing clouds.

To summarize: none of the techniques involved in Fourier-multiplex spectrometry is to be undertaken lightly and for this reason the description of Fourier spectrometers is curtailed here.

---

[6] The frequency spectrum is not to be confused with the infra-red spectrum: it refers to the frequencies (measured in Hz) present in the moment-to-moment variation of the source strength.

12
Detectors

In spectrography, where spatial resolution is required, there is a clear subdivision in detection technology between photochemical and photoelectric methods. So far as chemical methods are concerned they possess few advantages over photoelectric methods except possibly for observations at remote sites where a reliable supply of electricity is uncertain or unavailable. They comprise the reduction of silver halide salts to colloidal silver in a solid emulsion coated on to glass or on to a transparent flexible substrate.[1] The detective quantum efficiency (DQE) of silver halide crystals is in the region of 10^{-4}–10^{-3}, the dynamic range is in the region of 100:1 and as photodetectors they are severely non-linear in response, especially where long recording times – greater than 100 s – are involved. ISO ratings, usually quoted in the range of 12 for high-resolution fine-grain emulsions, to 2000 at the higher end of the range for severely coarse high-γ high-sensitivity emulsions,[2] fall rapidly at exposures longer than 10–30 s. The effect, known as 'reciprocity failure', is the notion that the blackening produced depends solely on the product of intensity and time and is a necessary assumption if spectrophotometry is to be done.

This sensitivity range may be compared with a liquid-nitrogen cooled CCD detector with a DQE in the region of 0.5 and strict linearity over a dynamic range of 25 000:1, a sensitivity broadly equivalent to an ISO of 3×10^6.[3] Liquid-nitrogen cooled CCD detectors require far more careful screening from light than silver halide emulsions. An ordinary photograph with a cooled CCD camera and an F/2 lens in a typical photographer's darkroom will give a clear, detailed picture of the interior in a few seconds' exposure.

---

[1] Gold salts can also be made to follow the same photoselective reduction process.
[2] Sometimes known as soot-and-whitewash emulsions.
[3] No strict equivalence is possible because of the non-linearities and different dynamic ranges involved, and this estimate is made by comparing the exposures at a shutter speed of 0.02 s needed for photographs of similar quality.

12.1 Silver halide photography

Emulsions on glass plates have become comparatively rare in spectrography and it is now the custom to rely on sheet film. Spectrographic plates have traditionally been large, 75 mm × 250 mm being typical, and the emulsions coating them have been specialised to cope with the peculiar requirements of spectrography. Chief among these is the need for uniform sensitivity across the whole spectral range covered by the instrument and generally instruments which deal chiefly with infra-red radiation are not required to cope simultaneously with the visible and UV regions.

12.1.1 Infra-red emulsions

Emulsions are available from photographic material manufacturers which will record radiation to ~1.2 µm in the infra-red. These have a comparatively coarse grain so that a spectrograph with high dispersion is needed for high-resolution spectrography. The roll film, sheet film or plates must be kept in a refrigerator until shortly before they are to be exposed, to avoid them fogging from thermal radiation. Condensation is then a problem especially in tropical regions, and the installation of a drying agent such as silica gel in the spectrograph is advisable *well before the plates are introduced*. Appropriate development conditions are required afterwards and the manufacturer's instructions must be followed closely for satisfactory results. If any sort of photometric precision is intended, the technique is an exacting and unforgiving one.

12.1.2 Visible and near UV emulsions

For the visible region and near UV, so-called 'astrographic' emulsions are available which were developed initially for astronomical research and which possess reasonably fine grain and uniform sensitivity throughout the visible region. The point which must be emphasised here is that if spectrophotometry is to be attempted, it is important that all the plates or sheet films come from the same batch, that they be developed and fixed under rigidly identical conditions of time, temperature and agitation, and preferably with the same batch and conditions of mixing of the developer. De-ionised water is recommended but is not of overriding importance.[4] This is provided that the water is free from hostile cations, i.e. that it has a pH higher than 7.

The bugbear of reciprocity failure at long exposure times has been the subject of a great deal of experiment, chiefly with the use of reducing agents to remove any

---

[4] The author was advised by a manufacturer that, on an eclipse expedition to a tropical island (Manuae, Cook Islands, 30 May 1965), their plates could if necessary be washed in sea-water, the final clearance washing being deferred until a later date.

trace of free oxygen. Hydrogen bathing and soaking in a solution of ammonia have been claimed to be effective, but following the development of the CCD, they are no longer worth considering seriously.

With experience, sufficient uniformity can be reached with the silver halides that spectrophotometry to 1% is possible. This seems to be a statistical limit, such that there is little improvement on it even with large amounts of repetitive data.

Emulsions can be sensitised to the far UV by suitable phosphorescent materials such as fluorescein, and sensitised emulsions are available commercially. For the extreme UV (XUV) below ~ 2000 Å Ilford 'Q' plates are to be preferred. These are manufactured, chiefly for the nuclear industry, without the clear top-coating of emulsion, so that the silver halide crystals actually reach the surface. Such emulsions, on thin glass plates, can be used in vacuum spectrographs and can be bent on a mandrel or in a plateholder to fit a 500 mm radius Rowland circle. The emulsion surface is sensitive to touch and so the plates require careful handling in the darkroom when they are loaded. As usual with silver halide emulsions, different grades are available with different crystal sizes and corresponding different sensitivities.

Included in the paraphernalia of traditional silver halide spectrophotometry must be a plate measuring machine or travelling microscope, and a densitometer is required if accurate photometry is to be done. Calibration exposures must also be made, although the method of the 'common calibration curve' is a useful alternative.[5]

12.2 Elementary electronic detectors

12.2.1 Photocells and photomultipliers

In the age of the monochromator the photomultiplier led the way for 50 years. It is a vacuum tube and comprises typically a photocathode deposited on an end-window – a thin film of silver-silver oxide overcoated with caesium, antimony or other substances with a low work-function, to act as a photoelectron emitting surface, and a succession of 'dynodes': electrodes with a slatted structure similar to a Venetian blind and held at successively increasing positive voltages to attract electrons sufficiently strongly to provoke secondary emission and thus to act as current amplifiers. A single photoelectron expelled from the photocathode will provoke a cascade through anything up to 13 successive dynodes until a pulse of $\sim 10^7$ electrons arrives at the anode. The pulse may have a duration of substantially less than a nanosecond[6] allowing high pulse rates to be collected. The pulses can

---

[5] See Appendix 4.
[6] Bearing in mind that the speed of light is ~ 1 foot per nanosecond, this is not so surprising.

be collected as a photocurrent or, especially if the radiation rate is very low, can be amplified and counted as individual pulses in a digital recording system. Either way, the DQE of the device is in the region of 0.15–0.35. Photomultipliers must be cooled, usually to $-80\,°C$ by dry-ice or a Peltier cooler to reduce thermal electron emission to a low enough level to allow faint light photometry.

The sensitivity of the photosensitive surface is wavelength dependent and there is little or no sensitivity at longer than 1.2 μm. Ultra-violet sensitivity can be improved by using quartz windows or by covering the glass with a thin layer of phosphorescent material – wavelength converters – typically fluorescein, vacuum-grease or sodium salicylate. Sensitivity to the phosphorescence from these materials continues down through the XUV to the soft X-ray region.

The chief sources of noise ('dark-counts') are thermal emission from the photo-cathode, cosmic ray Čerenkov emission in the glass and radioactivity in some of the materials, usually ^{40}K, in the glass.

There is an hour or two of high dark-current after the PMT has been exposed to light and this is due to phosphorescence in the glass.

12.3 Detectors with spatial resolution

12.3.1 The charge-coupled device

The detailed electronics need not concern us. It is sufficient here that a CCD is a rectangular plane surface covered with a two-dimensional array of 'pixels', each one a capacitor capable of holding up to half a million electrons; that the capacitor becomes charged with photoelectrons when light falls on it; and that the DQE, which is strongly wavelength dependent, is of the order of 0.5. The pixels can be 'read-out' (destructively) in sequence and the charge contained in each can be measured with an accuracy of $\sim \pm 5e$. The size of each rectangular pixel may be as large as 25 μm square (they are mostly but not necessarily square) and, at the time of writing, as small as 5 μm × 5 μm. The number on a chip, again at the time of writing, is as low as 80 × 80 for some applications, and with 9 μm pixels, as large as 4096 × 4096. Small pixel size is not necessarily an advantage in spectroscopy, bearing in mind that the efficiency of the spectrograph depends on the total *area* of the chip, not on the number of pixels it holds. The resolution afforded by a given size is described in the next section.

12.3.2 The dispersion relation

There is a fundamental relation, irrespective of the type of mounting or grating in use, between the spectral bandwidth $\delta\lambda$ falling on to one pixel of a CCD, the grating constant, a, the order of diffraction, n, the pixel width, p, the width, W, of the ruled area of the grating and the final camera focal ratio, F.

(1) From the grating equation we have $\cos r \, dr = n \, d\lambda / a$.
(2) The pixel width receiving the bandwidth $d\lambda$ is $f_2 \, dr$, where f_2 is the focal length of the output mirror (or lens).
(3) The projected width of the grating on to the output mirror is $W \cos r$.

The focal ratio of the output mirror is then $f_2 / W \cos r$ and $W \cos r \, dr / f_2$ is a Helmholtz–Lagrange invariant of the system which holds good even when there is re-imaging of the spectrum by a new output lens.

From these equations we can eliminate $\cos r$ and dr to give a universal relationship

$$\delta\lambda = pa/nWF. \tag{12.1}$$

This may well be taken as the starting point for the design of a CCD spectrograph.

One caveat: this relation holds only when the grating is the iris of the system. In the mounting described in Subsection 8.8.6 at the end of Chapter 8, for instance, it is the camera pupil diameter rather than the grating which determines W in Eq. (12.1).

12.4 Exposure limitations

The chief cause of problems with the cooled CCD detector comes from background, chiefly cosmic, radiation. At sea level, particles leave tracks or more often clumps of ~5–10 filled pixels. The diurnal variation is large but on average is ~50 CREs $cm^{-2} \, hr^{-1}$. These clumps, called CREs (cosmic ray events) can usually be recognised because they contain a much greater density of charge than normally exposed pixels. There is some dependence on the orientation of the CCD. It is no great matter to write a short program to detect such clusters in the read-out array and, if appropriate, change the affected pixel contents to zero. Due allowance can then be made in the reduction process – summing pixels along a spectrum line for instance – to disregard such empty pixels. The rate at sea level is such as to interfere substantially with spectrometric exposures much longer than 30 minutes, and a sequence of several exposures, median-filtered after read-out,[7] may be preferable to one long exposure.

Another, similar cause of blackened pixels is occasionally to be found in the lens of the CCD camera. Modern lenses with focal ratios of F/2 or less often contain

---

[7] The median is taken of the corresponding pixels in the sequence of exposures and a new photograph is compiled from all these chosen medians. This process is adequate to eliminate nearly all the CREs on the frame. Three frames is a minimum, four or five is better.

elements made from a glass which includes a thorium salt in its recipe. Such a glass is more than usually radioactive, not seriously enough to be a health hazard but enough to cause a rash of filled pixels on the CCD. A Geiger counter is a useful accessory when buying a CCD camera lens. Other possible sources of read-out noise are the materials of construction of the laboratory building housing the camera. A CCD photograph of a completely dark sub-basement room with unpainted brick walls will show the outline of the concrete between the bricks after about five minutes' exposure. This is from the radioactivity (chiefly ^{40}K) in the cement. Simultaneously it will be discovered that most materials in everyday use, including emulsion paints, are phosphorescent with decay times of up to several hours. This is a factor to consider seriously when choosing the materials of construction of the spectrograph. Test exposures with a newly constructed spectrograph with the entry slit completely closed may cause some dismay to the designer at first.

12.5 CCD software

Close inspection of the software attending a LN_2- or Peltier-cooled CCD camera is advisable. It is most likely that suites of programs have been written by the manufacturer to take care of spectroscopic applications, but there are certain items which are important if the best use is to be made of the camera. Most importantly, the whole of the raw data, that is, the two-dimensional array of pixel contents, must be available, pixel-by-pixel. All data processing must be accessible to the worker, and for this the FITS format is desirable.[8] Useful operations on the data include:

(1) All individual pixel contents must be available for inspection and alteration, preferably in a frame of about 8×8 pixels. Occasionally it is necessary to 'smooth' the data by replacing pixel contents with numbers in adjacent pixels. This is particularly so when CREs are obtrusive.
(2) There should be a graphic display of raw data either as a photograph or line-by-line in graphical form. Similar displays of processed data are also highly desirable.
(3) All arithmetic operations must be available, such as adding, subtracting, multiplying, dividing and logic operations on corresponding pixels in two or more frames.
(4) Median filtering of contiguous sets of frames is required.
(5) Binning of frames in the x- and y-directions separately or together by user-prescribed amounts should be possible, in order to alter the aspect ratio, with or without averaging the binned pixels. This includes binning down to a one-dimensional output to make spectrograms.
(6) Output should be available in ASCII format for further processing by external, probably spreadsheet programs.

---

[8] See Section 10.8.

(7) Rapid scanning of a chosen line comprising a few (20–30) pixels is useful for focusing the camera. Similarly, rapid scanning of a small frame of perhaps 50 × 50 pixels is useful for focusing and camera adjustment generally.

12.6 CCD calibration

The CCD sensitivity varies from one pixel to another, and from one exposure to another, and one may not rely on reproducibility to better than ~1%. Each CCD must be tested by itself for consistency and the results are sometimes disappointing. CCD photographs are generally divided by a frame which has (hopefully) been exposed to uniform illumination in order to eliminate errors due to pixel-to-pixel variation. This is sometimes called 'pre-whitening' or 'field flattening', although the latter is really an optical term.

This must be done through the camera with the same lens and at the same focal ratio as the observational photographs.

The pixel contents are then normalised to unity so that the division does not affect the overall exposure.

The chief problem lies with the phrase 'uniform illumination', which is more difficult to achieve than is generally realised. Plain white (i.e. high albedo) surfaces are rarely perfect Lambert radiators. Thick, translucent sheets of polythene are a reasonable approximation and a 75 mm radius hemisphere cut from PTFE, and made to cover the lens, is suitable for 35 mm camera lenses, provided it is illuminated by diffuse light. A straight-sided fish tank, filled with fresh water and half a litre of fresh milk, illuminated from a 10-metre distance by a small diameter source, will serve for larger apertures, but reflections are a problem and there must be complete darkness behind the camera lens. One may test for uniformity by making successive exposures with the optic axis at different angles to the surface normal of the glass.

The local variations of sensitivity will depend on the wavelength of the light and the resultant reference exposures are at best no more than an average.

12.7 Spectrograph calibration

There are so many variables in the optical path of a spectrograph that calibration is totally empirical. A black body at a known temperature is the only feasible method of absolute calibration, and even then one must be sure that there has been a total suppression of other orders and of scattered light in the spectrograph. Reliable *absolute* calibration is such an onerous task that one must question the need for it in most circumstances. For practical purposes a comparison with a standard source is sufficient. A reasonable approximation – no more – can be obtained by exposure to a card coated with freshly deposited magnesium oxide from burning magnesium ribbon and exposed to the midday sun, which in the visible at least, is an approximation to a black body at 6000 K. A white card will not do, neither will a painted white matt surface. Both are likely to be fluorescent under the solar UV radiation.

The resulting spectrum, after correction for the CCD area variations of sensitivity, may not look like a black body spectrum and for good reason. Apart from the basic variations with wavelength of the CCD overall sensitivity, there are the reflection coefficients of the various mirrors and, above all, the variations with wavelength and polarisation of the grating rulings, and the variation of these with angle of diffraction. Spectro-polarimetry with a reflection grating is probably a lost cause for this reason.

13
Auxiliary optics

13.1 Fore-optics

The correct use of a spectrograph requires that its aperture and its field be filled with light. More than this is not possible. Less than this is inefficient and may record the spectrum incompletely or incorrectly.

For most efficient use, the light source should be imaged on to the entry slit with a condensing lens, at a magnification which allows the whole slit length and width to be illuminated. The aperture of the condensing lens must be enough to allow the diverging light beam, after passing the slit, to fill the grating with light.[1]

When this is done the condensing lens must itself be imaged on to the grating (not the collimator) by a relay lens[2] somewhere near the entry slit, at a magnification sufficient to fill the grating with the image. If the condenser has been chosen properly in accordance with the first paragraph this will happen automatically. The aperture of the collimator must itself be large enough to avoid vignetting the grating, and the length of the entry slit and the distance of the collimator from the grating may well determine whether this is the case. The sort of fore-optical arrangement which meets these criteria is shown in Fig. 13.1.

However, there are perils in imaging a structured source on to the entry slit unless perhaps the source structure itself is being examined. In particular, in double-beam spectrometry, it is particularly important that the two beams see the same light distribution, and in this case the source should be imaged on to the grating[3] and not on to the entry slit(s).

---

[1] The use of 'lens' and 'grating' here is for economy. The words 'mirror' and 'prism' are implicit, as the principles are the same for all optical instruments.
[2] Also called a field-lens, or sometimes a 'Fabry' lens after its inventor.
[3] Or on to its image formed by a Fabry lens before the entry slit.

13.1 Fore-optics

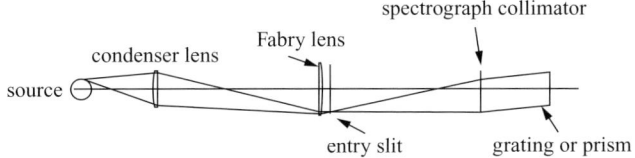

Figure 13.1 The standard method of illuminating a spectrograph. The essence of the arrangement is that both the field (the entry slit) and the aperture (the grating) are filled with light from the source. More is not possible; less is inefficient.

13.1.1 Telecentric operation

It sometimes is important when examining a remote extended source such as the night sky to be certain that all the line images receive their light from the same area of the source. To do this the source should be imaged on to the entry pupil of the instrument.

The simplest way to ensure this is to place the input (M1) mirror of a Čzerny–Turner spectrograph at its focal length from the grating. The chief-rays from bundles from all parts of the entry slit are then necessarily parallel to the optic axis when they reach the grating, so that all ray bundles have come from the same part of the distant source. The source is itself focused on to the grating which itself acts as a telecentric stop.

The input mirror may thus be at a different distance from the grating from the output M2 mirror which, if it is a paraboloid, should be at two-thirds of its focal length from the grating to ensure a flat field.

The alternative way of ensuring telecentric input is to put a Fabry lens at or near the entry slit, its focal length chosen so that together with the M1 mirror it images the distant source on to the grating. If the M1 mirror is also at the two-thirds focal length position, this may need a very long – possibly impractically long – focal length in the Fabry lens.

13.1.2 The telecentric stop

'Telecentric' is a technical term taken from the discipline of profile projection and cathetometry. It implies that chief-rays of ray bundles from the object are all parallel to the optic axis, so that a true silhouette of an object can be photographed.

In spectrography it may happen that an active beam of particles – a beam of excited molecules for instance – is to be examined and that a spectrum is desired with the beam imaged on to the entry slit, perhaps to measure the decay rates of various transitions. It is important then that there should be telecentric illumination

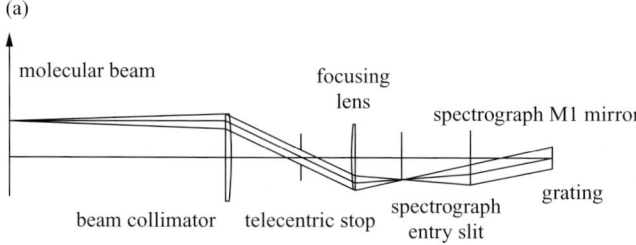

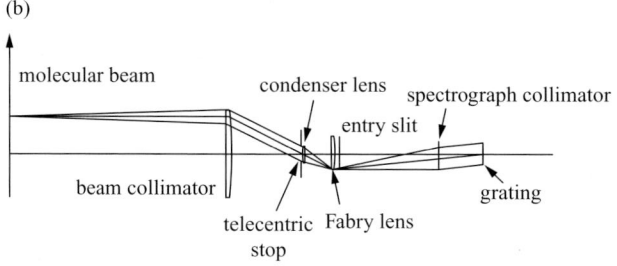

Figure 13.2 Two methods of achieving telecentric input to a spectrograph. Precautions like this are necessary for avoiding Doppler effects for example, when examining phosphorescent decay times in molecular beams. All chief-rays are perpendicular to the beam direction.

(in the object space only) in order that errors of wavelength due to the Doppler effect are avoided.

For this a lens or mirror is necessary **with a diameter greater than the length of beam to be examined**.

This lens acts as a collimator and at its rear focus is placed a stop, the 'telecentric stop', which ensures that all ray bundles have their chief-rays parallel to the optic axis. At or near this stop is placed the condenser lens which focuses an image of the beam on to the entry slit. A Fabry lens in front of the slit images the condenser on to the grating as shown in Fig. 13.2. Alternative arrangements are possible but this one allows the smallest diameter lenses to be used.

If accurate imaging of the beam is required, astigmatism must be avoided and it may be necessary to use Littrow optics to achieve this.

13.2 The astronomical telescope as fore-optics

If the spectrograph is to be attached to an astronomical telescope, then in the ideal case the star would be focused on to the entry slit and the telescope objective simultaneously imaged on to the grating. This however is a doctrine of perfection. The chief problem is that even if the imaging is perfect the diffraction-limited star image is far too large for the spectrograph entry slit. If the spectrograph input collimator is made long enough for the stellar image to fit the slit width, the image of the telescope objective is far too large for the grating. Neglecting atmospheric 'seeing' turbulence, the spectrograph used at its full resolving power should have an entry slit width such that the principal lobe of its diffraction pattern just fills the grating. For perfect efficiency therefore, the grating would have to be the same size as the telescope objective when the diffraction-limited image of a star falls on to the slit. Therein lies the essence of the problem.

Many ingenious solutions have been tried with varying degrees of success, none of them overwhelming. The 'image slicer' family takes sections of the stellar 'seeing disc' on either side of the slit aperture and redirects the light through a more or less complicated set of mirrors to a part of the slit above or below the seeing disc.

Specially formed fibre optic bundles offer a more reasonable solution to the problem.[4] For example, at the Cassegrain focus of a 4-metre telescope, working at F/12.5, a 0.5 arcsecond seeing disc – the smallest one may reasonably expect at or near the bottom of the atmosphere – has a diameter of 125 μm. An F/12.5 spectrograph with a 200 mm wide grating at a resolution of 10^5 requires an entry slit width of 12.5 μm, and a focal reducer to convert the F/3 output beam from the fibre to the F/12.5 of the spectrograph must start with an object diameter – the end of the fibre – of 3 μm. Therefore ~1800 such fibres are required to collect the light from the whole of the seeing disc and will convert it to a line 3 μm wide and 5.4 mm long. This becomes, on magnification, 12.5 μm wide and 22.5 mm long, about as long a slit as the spectrograph can tolerate without unacceptable aberration.

13.2.1 Slitless spectrography

The stellar astronomer has the advantage that the light sources to be investigated are already collimated and are (optically) at $-\infty$, and the function of the telescope is to collect as much light as its aperture will allow while retaining the collimation. The problem has already been pointed out, that the stellar image, even when not broadened by atmospheric turbulence, is too large to allow efficient high-resolution spectrography. The chief advantage of slitless spectrography therefore is in its use with low magnification, or no magnification at all, to examine simultaneously a large number of sources, each of which will record its own spectrum on a different part of the field. The application is therefore in taxonomy – the measurement of the relative abundance of different stellar spectral types for example, or the relative

---

[4] Fibre bundles with individual fibres a few micrometres in diameter may have a circular form at one end to receive the starlight, and be shaped to a single column of fibres at the other end to match the input slit.

abundance of galactic spectral types – Seyferts for example – in a cluster: this latter at high magnification of course.

The spectrograph input optics then are similar to those of the eye: the field of stars at the telescope focus is re-collimated by the 'eye' lens, and the 'field' lens[5] images the telescope objective on to the grating. The output optics of the spectrograph focuses the star field on to the photographic plate or CCD detector. In the latter case the focal reducer, described in the next section, may be necessary.

High resolution is not to be expected in slitless spectrography, spectrograph aberrations being what they are, but enough information can be gathered by this technique to make it a valuable asset in astrophysics.

13.3 Focal reducers

The focal reducer was invented originally by Meinel[6] and separately by Courtès[7] as a means of increasing the relative aperture of an astronomical telescope. In effect it converts a 5-metre telescope into a 35 mm camera with a 10-metre, F/2 lens.

The principle is that of a giant eyepiece. Following the Cassegrain (or any other) focus, the paraxial beam is allowed to expand until it has the same diameter as the final camera pupil. It is then collimated with a simple achromat working at the same focal ratio, typically F/14, as the Cassegrain telescope. Simultaneously a field lens just beyond the Cassegrain focus images the telescope objective on to the camera entry pupil.

The device can be further simplified.[8] Using a modern camera lens with a 'macro' facility, one may dispense with the collimator, and a simple field lens of appropriate focal length is the only addition required. The depth of focus is such that it need not be achromatised and a simple plano-convex lens is adequate. The camera pupil is placed at the image of the telescope objective.

The device may usefully be applied to spectrography, especially for the study of faint sources. A properly designed focal reducer will convert a Čzerny–Turner CCD spectrograph to a focal ratio of F/2 or thereabouts without loss of resolution. The principle is the same: a collimator captures the beam expanding after the primary focus, at a diameter equal to that of the final camera entry pupil. Simultaneously a field lens just beyond the Čzerny–Turner focus images the grating on to the camera entry pupil. The focal reducer then designs itself, being constrained by these two requirements. The design should be that of a simple Ramsden eyepiece and both field lens and eye lens are plano-convex with the convex surfaces facing each other.

---

[5] These words being used by analogy with a telescope eyepiece.
[6] A. B. Meinel, *Astrophys. J.*, **124** (1956), 652.
[7] G. Courtès, *Ann. d'Astrophys.*, **23** (1960), 115.
[8] J. F. James, *Q. J. R. Astron. Soc.*, **22** (1981), 244.

13.3 Focal reducers

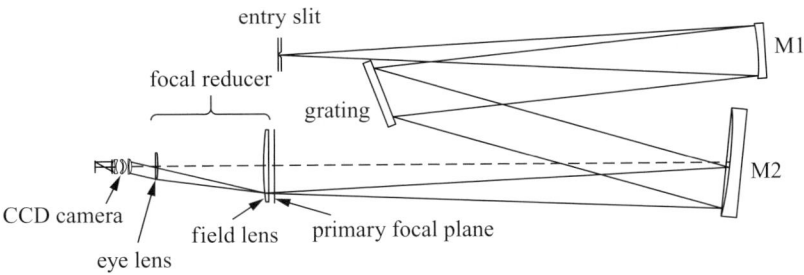

Figure 13.3 A Čzerny–Turner spectrograph with a focal reducer. The device achieves a low focal ratio without loss of resolution, but cannot be added to an existing design. The Čzerny–Turner spectrograph itself must be designed to have a flat primary focal surface perpendicular to the zero-coma chief-ray, and this is a matter for experiment with a ray-tracing program. A simple Ramsden-type reducer then adapts the spectrum to the CCD surface while the grating is simultaneously imaged on to the camera pupil.

As Conrady pointed out, attempts to improve the resolution by bending either lens are doomed to failure or, in his own elegant phrase, 'do not open any very alluring prospects'.[9] The field lens focal length is chosen to image the grating on to the camera entry pupil at a distance of $[(n-1)/n]f_{\text{eye}}$ beyond the eye lens, at the position that is, of the coma-correcting stop for the plano-convex eye lens. An example of the arrangement is shown in Fig. 13.3.

The Ramsden type of eyepiece is to be preferred because the Čzerny–Turner focus is accessible so that, if necessary, stops can be placed there to intercept line images which are overwhelmingly bright and scatter unwanted light into the camera. A fluorescent screen can also be placed there for observing UV spectra beyond the scope of the detector. (Such a screen may be of plane glass coated with a thin transparent film of a fluorescent material.)

A disadvantage is that this simple focal reducer will require refocusing if a different part of the spectrum is to be examined, and the eye lens may need to be an achromatic doublet to avoid this. The proper position for the camera entry pupil may be calculated or, more likely, will be found by experiment with the ray-tracing program. There is no sensible contribution to the chromatic aberration by the field lens which may remain a simple plano-convex element.

A small amount of chromatic aberration can be deliberately introduced to the spectrograph itself, by moving the entry slit inside or outside its true focus. This causes the grating to be illuminated by slightly convergent or divergent light and the convergence after diffraction is diffraction-angle, i.e. wavelength, dependent via the $\cos i / \cos r$ factor in Eq. (8.10). The final focus then depends on the wavelength,

---
[9] A. E. Conrady, *Applied Optics and Optical Design* (London: Oxford University Press, 1929), Vol. 1, 482.

and the slit position shift – a few mm – is small enough that there is no appreciable alteration in the other aberrations.

A 1-metre Čzerny–Turner spectrograph with a grating of 100 mm ruled width and 12 000 rulings/cm will spread an 800 Å band of the visible region across 110 mm at F/10 at the primary focus, and the focal reducer will convert this to 26 mm at F/2, covering a 25.6 mm CCD with 1152 pixels at a resolution of 0.7 Å.

13.4 Schmidt-camera spectrography

The Schmidt camera is a well-known optical device for achieving a very high numerical aperture at a large field angle. It has been applied to spectrography for the observation of very faint extended sources such as the zodiacal light and the night airglow. The input optics is conventional, with a long-focus collimating mirror or lens and a plane grating, but with the output optics replaced by a Schmidt camera. The property to be used is the high numerical aperture, and no great field angle is required. With a grating constant of 6000 rulings/cm the visible spectrum is covered in first order by a field of $\pm 3.5°$.

Briefly, the Schmidt camera uses the condition for zero coma in a spherical mirror given by Eq. (4.8): the pupil must be at $z = -2f$, that is, at the centre of curvature of the mirror. At this pupil a zero-power corrector plate figured to a fourth-order surface curvature is placed to remove the spherical aberration, which at this high numerical aperture would be severe. Symmetry precludes astigmatism (Eq. 4.9) but field curvature must be accepted as the price to pay for removal of the other aberrations at such a high numerical aperture, and herein lies the problem with spectrography. In a large Schmidt camera a photographic plate can be bent over a mandrel to accommodate the field curvature and in a small one the emulsion can be coated on to a spherical glass substrate; or a field-flattening optical system can be placed in the appropriate position. But a CCD detector requires such a bulk of electronics and refrigeration that a totally unacceptable obstruction to the beam would be incurred and it is this which makes the Schmidt configuration impracticable for CCD spectrography. A high numerical aperture depends therefore on a focal reducer followed by a good quality camera lens of low focal ratio.[10]

13.5 Scattered light and baffling

The purpose of including baffling in the design is the reduction of scattered light.

---

[10] For most spectrographic purposes a standard double-Gauss 35 mm F/2 camera lens, of which there are probably a hundred designs on the market, is adequate.

13.5 Scattered light and baffling

13.5.1 Causes of scattering

Scattering comes chiefly from uneven edges to the input slit, from dust and scratches on any of the optical surfaces, from imperfections in the grating rulings and from light coming through the entry slit which reaches any of the walls of the interior of the spectrograph. Baffles are simply opaque materials in the form of walls or tubes which intercept and absorb unwanted light. They may be inside the spectrograph or outside or both.

Deficient optics are of course a matter of quality control by the manufacturer, and large, thick lenses of glass or quartz are more likely to have striae, stones, sleaks and scratches than small ones. Mirrors are superior in this respect but still must be manufactured, coated and chosen with care. Dust particles on optical surfaces are a matter of cleanliness and accessibility. Gratings, by the nature of their construction, scatter more than most optical elements and for this reason prisms are still to be considered when scattered light is more of an obstacle to research than lack of high resolution.

13.5.2 Scattering from edges

The edges of the entry slit were mentioned above. Any edge which can be seen from the grating is suspect, because of the possibility of Fresnel diffraction effects. The sharp edge of a lens-hood is a particular culprit even in a camera, and may scatter enough light on to the lens surface to be re-scattered into the interior. For true baffling, several such edges must shield the lens, each one shielding from the last: alternatively, a trumpet-shaped baffle will show a smooth horizon rather than a sharp edge to the input optics. The problem is particularly acute in space-borne spectrographs where it is desired to acquire stellar spectra from anywhere within 90° of the Sun.

13.5.3 External baffles

The principle to follow when judging where to incorporate baffles is that the grating must 'see' nothing except light from the external condenser, and that light entering through the slit can go nowhere except to the grating.

An external baffle in addition to the internal baffle may help to achieve this. There is an image of the grating formed by the collimating mirror and the input Fabry lens. It is situated *before* the entry-slit/Fabry-lens assembly. There may well be a condensing lens there, but whether there is or no, this is the place to put a real stop, its aperture cut to the shape of the grating, inserted in the optical train and tilted to match the image of the tilted grating. The tilt is not large, bearing in mind that the longitudinal demagnification is the square of the lateral demagnification of the grating image which it covers. It must follow then that light which enters the spectrograph through this stop, unless scattered by the input slit, must fall on to the grating. Similarly, if there is a re-imaging of the spectrum, for example through a focal reducer, there may be another opportunity to insert a made-to-measure stop at the image of the grating on the output side, although this will coincide with the camera entry pupil.

13.6 Absorption cells

Absorption spectra of liquids and solids require only the simplest of absorption cells, typically with paths from a few micrometres to a few centimetres, and such cells are normally placed just before the entry slit, sometimes covering only one half of it so that comparison spectra come through the other half. If this method is adopted, then the designer must be careful to see that the whole slit length receives identical illumination and it is the light source and not the condenser which must be imaged on to the grating by the Fabry lens. Strictly speaking, the light should be collimated, but in practice a low numerical aperture, equivalent to $\sim F/12$, is adequate to give the accuracy normally required. The marginal rays in the cell differ in length by less than one part in a thousand from the paraxial ray.

13.6.1 White cells

Serious problems only begin to arise when the absorption is very low so that long paths are required.

In the White cell,[11] long path absorption is achieved through the periscopic principle, in which a beam of light is sent down a tube with relays of lenses which achieve a useful field of view while confining the light to the interior. In an absorption cell the lenses are replaced by mirrors, and absorption path-lengths of tens of metres can be achieved in a 1-metre long tube.

Three concave spherical mirrors are employed, all of the same radius of curvature (Fig. 13.4). Light is admitted through an entry slit and is reflected off mirror 1 to a focus at a point on the surface of mirror 2. From mirror 2 it is returned to the centre of mirror 3 and thence to a focus at another point on mirror 2. The process continues until a final focus is reached outside mirror 2, whence it goes to the spectrograph. Figure 13.4 shows the chief-ray as it passes through the cell with seven reflections.

---

[11] J. V. White, *J. Opt. Soc. Am.*, **32** (1942), 285.

13.7 Fibre optical input

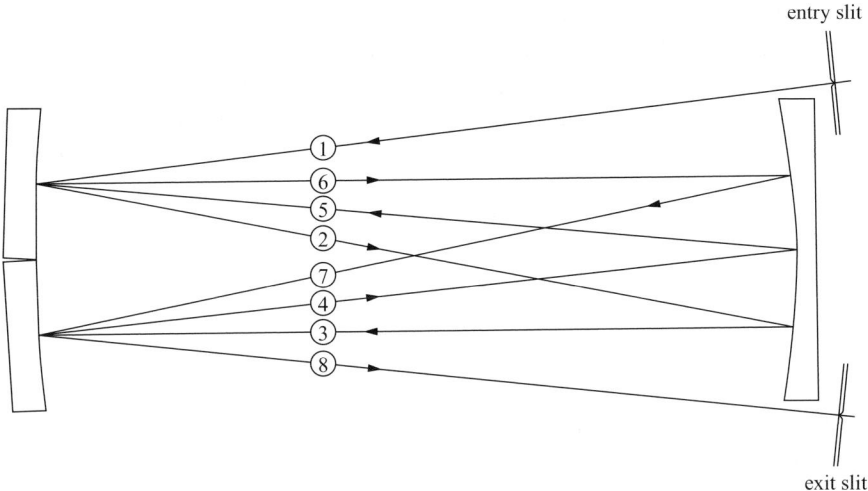

Figure 13.4 The White cell. This is a device for measuring feeble absorption coefficients where long path-lengths are required. The device uses multiple reflections between three mirrors and depending on the reflectivity of the mirrors, the number of reflections may be as high as 30 in the visible and several hundred in the FIR where reflection coefficients are higher.

The actual number will depend on the tilt of the two smaller mirrors, which may be adjustable.

The number of reflections available depends on the reflection coefficients of the mirror coatings. An untarnished silver coating has a reflection coefficient in the region of 0.92 and transmits 10% of the incident light after 27 reflections. Higher reflection coefficients and longer paths are possible over restricted wavelength ranges with multilayer coatings. Gold coatings in the infra-red, with reflection coefficients in the region of 0.98–0.99, will allow several hundred reflections.

13.7 Fibre optical input

The salient facts which concern us in spectrography are that multiple fibre bundles are used rather than single fibres, and that a circular luminous object, such as a star image, can be converted to a line image suitable for the input slit of a spectrograph. A single fibre in a bundle may have a diameter anywhere from 1 μm to 25 μm, 10 μm being typical, and the output beam from the bundle emerges at a focal ratio in the region of F/3. The illumination is Gaussian rather than uniform, but should be converted by a lens to ∼F/12 so that all the emerging light will be accepted by the collimator. The apparent slit width will then be four times greater than the diameter of the fibres at the output end and the apparent slit length will also be four times greater than the length of the output end of the fibre. A low-power microscope objective may be a suitable converter, focusing the fibre-bundle end on to the entry slit. A Fabry lens at the slit will be needed to image the converter pupil on to the grating. Alternatively a virtual magnified image of the fibre end may act as the entry slit, in which case aplanatic lenses are indicated.

14
Optical design

Le mieux est l'ennemi du bien.
(The best is the enemy of the good.)
Voltaire

14.1 First steps

Once the task for the spectrograph has been defined, a suitable type may be chosen and a catalogue search made for a possible manufactured model. Many factors affect the decision. The first of course is fitness for purpose. Time and cost of manufacture may be factors and the facilities available for local or in-house design and construction are primary considerations. A decision on whether to buy or to design and build a dedicated instrument must rest on such an appreciation.

It is a mistake when doing fundamental or academic research to construct a more elaborate, high-performance or expensive instrument than the immediate task demands, possibly in the hope or expectation that it may prove useful for some other investigation at some later date. In the author's experience this is almost never the case and the chief result is delay and unnecessary expense. In the long term the instrument is a sad relic, cannibalised of its optical components and left to decay in the attic with its ingenious mechanisms and precision micrometers.

14.2 Initial layout

Once the spectrograph type has been chosen and its main parameters such as wavelength range, resolution, type of detector, number of resolved elements, étendue etc. have been decided, a sketch can be made, and the traditional back-of-an-envelope is as good a place as any for this. Manufactured mirrors and lenses must be considered first before taking a decision to have optical parts made, and it may well be discovered that the design can be adapted to use bought-in elements. Almost certainly

140   *Optical design*

the optical components will come from a maker's catalogue. Rarely a bespoke item must be ordered from a specialist manufacturer and if this is so, optical manufacturers are usually happy to let the designer have lists of their tool radii, of suitable glasses at their disposal, with their optical properties and the costs of manufacture. Catalogues of optical, opto-mechanical and opto-electronic components must be available to the designer, quite probably from the Internet.

14.3 The drawing board

The next stage is to compute the Gaussian optics and to make the outline optical drawing, and this will determine the sizes and positions of the optical components.

When making the first sketch it is both clearer and more instructive to lay out all optical components in position as if they were transmitting light rather than reflecting it. Consider all the mirrors as if they were lenses and the grating as if it were a transmission grating.

The x- and y-scales on the diagram need not be the same. The ray paths are not affected and the positions of stops and pupils will be unaltered. A CAD (computer aided design) program may lend some precision to the drawing, but the old ruler, pencil and compasses is probably adequate and may well be more convenient at this stage.

If the instrument is to be of the Čzerny–Turner or Ebert–Fastie type, this is the time to choose input- and output-mirror off-axis angles for the chief-rays. It is a simple matter to write a short computer program which, given the zero-coma wavelength and the sum of the two off-axis angles, will compute, on the assumption of parallel input and output chief-rays, the α and β angles, the grating tilt and the meridional coma for rays on either side of the ZMC direction. If the instrument is to have a Littrow configuration, the off-axis angle should be chosen so as to allow adequate spacing for the various components with room for the mirror and grating cells, auxiliary mirror mounts, slits and CCD/plateholders. A computer ray trace will determine whether the aberrations are under adequate control. Assume at this stage that all mirrors will have a thickness equal to one-sixth of the diameter, and that all positive lenses will have a rim thickness of at least 2 mm. Grating blank sizes are to be found in the manufacturer's catalogue.

14.4 Computer ray tracing

This is the point to emphasise once again that one cannot design an optical instrument on a computer with a ray-tracing program.

The design must be done first and set up on the ray tracer for refinements to be made. Remember that the Seidel aberration calculations are not exact and that small changes in the dispositions of the optical surfaces can be made by the program to minimise the aberrations, reducing them sometimes to zero.

A typical ray-tracing program invites the operator to insert the x-, y- and z-coordinates of the vertex of each *surface*. The z-axis is normally the optic axis. It then requires the pitch and tilt angles of the surface, its curvature and excentricity, whether it is reflecting or transmitting, whether it is a grating and if so the number of rulings per millimetre. Refractive indices will generally follow each vertex, referring to the material following.

Next, the x-, y- and z-coordinates of the 'launch points' of each ray are required together with the sines of the angles which each one makes with the z-axis in the x–z and y–z planes, and the wavelength associated with the ray.

Bear in mind that the outer 3 mm at the rim of a mirror cannot be relied on, as there is more often than not a 'turn-down' as the result of the polishing action. If the manufacturer is supplying a 150 mm diameter mirror, it is prudent to assume that the aperture is 140 mm when doing the optical design. In any case the retaining plate which holds the mirror in its cell may well have such an overlap at the edge.

The program will trace the ray through the system with an accuracy far beyond anything that can be measured in practice. It can be asked to adjust any (or all) of the input parameters until the ray arrives at a definite given goal, usually either the x- and y-coordinates of a point on the last surface or, if the output is to be a collimated beam ready for a bought-in camera lens, the sines of the angles at which the ray leaves the last surface. A multiplicity of rays, 20 or 30 perhaps, can be treated simultaneously.

14.5 Refinement of the optical design

Thus, when the outline design appears satisfactory – sufficient room left for baffles, detector optics and fore-optics unlikely to interfere with each other etc. – the optical train can be set up in the ray-tracing program as outlined above. This is the design stage which formerly was done by hand using seven-figure logarithm and trigonometrical tables, when an experienced ray tracer could trace a ray through eight or nine surfaces in 20 minutes and an experienced lens designer could make an intelligent guess as to which parameter needed changing next before repeating the process. A modern ray-tracing program will trace 30 rays through 20 surfaces in as long as it takes to press the start key on the computer keyboard.

There is a sufficient variety of these ray-tracing programs on the market that it is pointless to particularise too much. No great elaboration is required here, as focal ratios are unlikely to be below $\sim F/8$, the limit at which simple achromatic doublets provide adequate field and resolution for a spectrograph. One aims generally at a spatial resolution in the neighbourhood of 3–5 µm depending on the type of detector and the resolving power required. A simple geometrical ray-tracing program such as 'BEAM 3' from Stellar Software[1] is adequate although it is strictly geometrical and does not include refinements such as diffraction effects at the image points.

In the event that a manufactured camera lens is to be used for the camera,[2] then the spectrograph designer's aim may be for parallel ray bundles at each computed wavelength at the final output pupil of the spectrograph.

14.5.1 Initial alignment

Once the optical elements are put in their places, a single ray, the 'chief-ray', at the central wavelength is launched from the centre of the entry slit to the vertex of the first (collimating) mirror or lens. The mirror is given the tilt required by the outline design and the grating (or prism first surface) position may be adjusted by the program to receive the ray at its centre.

The grating tilt will already be known from the outline design and this will have already been entered in the ray-tracing program. Alternatively, if the zero-coma wavelength has been decided, the grating can be tilted by the program to reflect the chosen wavelength to the vertex of the next surface.

The position of the second mirror is then adjusted to receive the diffracted chief-ray at its vertex and its tilt is then adjusted by the program to send the ray either towards the detector centre or possibly to be made parallel to the input ray. In the latter case the detector vertex may be adjusted – by the program – to receive the chief-ray.

Exact positions are not set firmly at this stage. The distance from the entry slit to the M1 mirror of a Čzerny–Turner mounting is found by tracing parallel rays *backwards*, starting from the corners of the grating and finding the place where the marginal rays would meet to form a comatic image of a virtual point at infinity. This is where the entry slit must go, and the procedure takes care of the field curvature of the M1 mirror and avoids any focus defect. Rays launched from a point source at the slit centre towards the M1 mirror then arrive at the grating with the proper amount of coma.

Procedures like this set out the initial dispositions of the major components, and ray bundles of six or eight marginal rays may be launched from the same point on the entry slit and adjusted (again by the program, which can adjust the sines of the launched rays) to arrive initially at eight equally spaced points around the margin of the iris of the system – usually the grating. These are then traced on through the system so that the final image quality can be assessed.

[1] Stellar Software, PO Box 10183, Berkeley CA 94709, USA.
[2] Recall the advice in Section 12.4 regarding radioactive elements in low focal-ratio camera lenses for CCD cameras.

This is the stage where such items as coma, astigmatism and field curvature can actually be measured (as opposed to being calculated) and necessary adjustments made to the positions and tilts of the various components to ensure compliance with the original design goals.

The process is a lengthy one and experiments – on the computer – are generally needed to find the optimum positions. The departures from the positions calculated by Seidel theory will not be large – a small percentage probably – but they will be significant, and the final aberrations may well be substantially less than the original estimates.

14.5.2 Elaboration of the design

Once optimum positions have been found for imaging a point source on the entry slit, the length of the slit can be added as a new condition and the effects of spectrum line curvature can be measured. It is at this stage that compromises will be needed to secure the best image and to discover the spectral range over which the whole spectrograph will operate satisfactorily.

When the process is complete and a satisfactory optical design has been reached, small changes should be made in the ray-tracing program to the positions on the optical layout, to determine the tolerances that may be allowed in positioning the various components. This in turn determines the accuracy which must be required in the workshop when the instrument is finally constructed. More importantly, it may well reveal that the manufacturing tolerances are too tight and that the design cannot be realised in practice. This sort of defect occurs, for example, when a ray is incident on a refracting surface at a large angle, and where a small displacement of the surface vertex along the optic axis would entail a large change in the height of incidence and consequently the ruin of all that follows. Considerations like this play a significant part in the design of camera lenses for example.

14.5.3 Local minima

A warning is appropriate at this point – a general warning in fact to everyone using a ray-tracing program. The program varies the positions, tilts, curvatures etc. as it searches for a minimum departure from the goals set by the programmer – a focus perhaps or parallelism of rays in a collimated beam. The initial fixed points of the program may be the launch points of the rays and the refractive indices of the transmitting elements. The variables are the bendings, excentricities, vertex positions, tilt and pitch of some or all of the optical components (some, taken from a manufacturer's catalogue, may be taken as fixed, except in position). Vertices can be ganged together so that a lens or a group of lenses can be moved together as one unit.

It is inevitable, with such a multiplicity of variables, even when only two or three are allowed to vary at one time, that many local minima will be found such that any small change in any one of the variables will produce a worse result. In practice a great deal of experiment will be needed to find the ultimate 'best' arrangement of the components, and the final result, which may sometimes be surprising, must be tested with a large variety of possible ray-input parameters.

Metaphorically, the climber is climbing a mountain in a fog, and must be certain that she has actually climbed the mountain and is not standing on the summit of one of the foothills.

The design found by computing the Seidel aberrations will almost certainly need minor corrections. The 'flat-field' condition of Eq. (4.12) will not give an exact position for the grating and a small movement of the grating or the M2 mirror, of several per cent of the M2 focal length, may be required to produce a field flat enough for the CCD detector. Computer experiment is the answer, and it is instructive to plot, for example, a rough graph of the way the field curvature actually changes with M2 position.

14.6 Requirements of a ray-tracing program

The obvious function of a ray-tracing program is to allow you to make small variations in such variables as you choose and to re-compute the rays to see whether the variation moves them towards or away from a goal which you have chosen. Several variables may be chosen at one time and this is the chief cause of the program's finding local minima.

There are other useful functions of the program beyond this. It should be possible to compute spot diagrams at the final focus or at any other intermediate point in order to check the contributions of the various aberrations. In a CCD spectrograph a square or rectangular iris at the position of a pixel or a column of pixels should measure the proportion of randomly launched rays which enter the target area, a potent measure of the efficiency of the optical design.

There are other, more subtle variables which can usefully be displayed. Apart from drawing the ray diagram, an x–y plotting program should be available to make graphs of one variable against another. For instance the so-called H–$\tan U'$ plot,[3] in which an oblique parallel bundle of rays at some angle with the optic axis is traced and a graph made of the intersection height, H, above the z-axis at the image plane, against some other variable such as $\tan U'$, the tangent of the angle at which each ray emerges from the exit pupil, or H, the intersection height of each ray at the entry pupil on the y-axis. A perfect lens would show the plot as a straight

---

[3] R. Kingslake, *Lens Design Fundamentals* (New York: Academic Press, 1978), p. 144.

horizontal line. The ray-tracing program will show, in the shape of the actual curve that is drawn, the effect of various aberrations of the system. A straight line with a slope indicates field curvature. An S-shape indicates spherical aberration. U- and V-shaped lines indicate positive and negative coma respectively, and if the curve is asymmetrical it indicates spherical aberration as well. Details of this technique are to be found in Kingslake's book.

There is a world of difference between lens design and instrument design. Nearly all spectrographs work at focal ratios greater than F/12 and field angles less than 3°, the points at which serious lens design begins. Seidel theory is adequate to give the iterative ray tracer a place to begin, and when low focal ratios are needed, photographic lenses can be made to look at the collimated beams which emerge from F/12 optical systems. The spectrograph designer will rarely have to deal with anything more complex than the design of an air-spaced achromatic doublet. It is the proper positioning of simple elements along the optic axis which chiefly gives refinement, resulting eventually in a high resolution and étendue over a wide spectral range.

14.6.1 Worked example

This is a 1-metre focal length Čzerny–Turner CCD spectrograph intended for night airglow spectroscopy, with angles chosen to remove tangential coma at 5000 Å in second order. The spectrograph was intended to photograph the spectrum between 4000 Å and 6000 Å in tranches of 500 Å at a resolution of 0.5 Å, turning the grating for each new spectrum and moving the focal reducer on a sliding carriage if necessary to refocus.

A Ramsden-type focal reducer was added to put the spectrum on to a CCD with 1152×325 pixels of 25 μm width covering 28.8 mm. The plano-convex lenses are of BK7 glass and have focal lengths of 497 mm (field lens) and 429 mm (eye lens). The focal reducer requires a longitudinal movement of a few millimetres to bring all sections of the visible region into focus. As it is a tranche of ~ 500 Å can be photographed at any one time.

It was designed using the 'BEAM 3' program mentioned earlier.

Table 14.1 shows the layout of the ray-tracing program for the Čzerny–Turner spectrograph with a focal reducer made to order, taking surface radii from the manufacturer's catalogue. The specification was that the grating ruled width was 120 mm, the (paraboloidal) mirror focal lengths were 1000 mm and the final CCD camera lens was to be an F/1.8 50 mm focal length 35 mm camera lens. The design requirement then was for parallel ray bundles to emerge from the focal reducer with a diameter of 25 mm.

Table 14.1

```
8 surfaces                              cupcztu
index   Z0              X0          M  C            Pitch       Dx  Dy   Gx   O   S   F
------- --------------- ----------- -- ------------ ----------- --- ---- ---- --- --- ---
   1    :1000.0        :0         :M :-.000505    : 2.948     :150 :150 :    :0  :   :r :
   1    : 180.168      :-84.664   :G :0           : 34.305    :122 :122 :1.2 :-2 :   :s :
   1    : 860.0        :-230.413  :M :-.000504    :-6.050232  :220 :220 :    :0  :   :r :
1.51872 :-181.183761   :-230.413  :L :0           :0          :195 :195 :    :   :1  :r :
   1    :-202.683761   :-230.413  :L :0.00392975  :0          :195 :195 :    :   :1  :r :
1.51872 :-545.740078   :-230.413  :L :-0.0045206  :0          :93.2:93.2:    :   :1  :r :
   1    :-553.740078   :-230.413  :L :0           :0          :93.2:93.2:    :   :1  :r :
   1    :-647.622278   :-230.413  :F :0           :0          :120 :120 :    :   :1  :r :
```

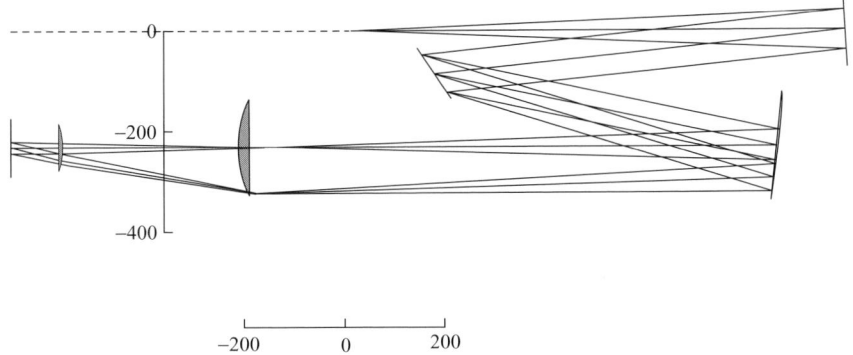

Figure 14.1 The optical layout of the Čzerny–Turner spectrograph described in the text. The focal reducer is a simple Ramsden-eyepiece design, that is to say, with plano-convex lenses having the plane surfaces outwards. This serves to collimate the emerging spectrum at a diameter of 23.4 mm and simultaneously images the grating on to the final image-plane where the pupil of the CCD camera lens would be placed. The camera lens (not shown) may typically be a 35 mm camera lens with a focal ratio of F/2 or less, and its optical quality is likely to determine the ultimate resolving power of the instrument.

The initial layout had the angles appropriate to zero coma at 5000 Å, with the mirrors at $z = 1000$ and the grating at $2f/3$ from the M2 mirror vertex, as required by the Seidel flat-field condition. Rays were launched from $z = 2.65$ for minimum coma at the M1 mirror, and the final positions for M2 and the following optics were achieved by the requirement for a flat field at the CCD, that is, for the emerging ray bundles at 5000 Å (paraxial) and 5250 Å (marginal) to be parallel, and in particular to have the x-direction cosines all as closely as possible equal to each other. The y-direction cosines – which describe the astigmatism – were not subjected to the iteration process. The question marks in the delimiter columns following the lens vertices were instructions to the iterator to vary the vertex positions of the focal reducer to minimise the aberrations. The letters d and f, in the column below each, were instructions to move the second lens surface with the first to keep the lens thickness constant.

Figure 14.1 shows the layout of the resulting spectrograph. In the final version the ray bundles have diameters of 23.4 mm. This compromise came from the best choice of available tool radii when the lenses for the focal reducer were ordered.

Figure 14.2a shows the spot diagrams of the final collimated output beams at 5000.00 Å and 5000.15 Å, showing the eventual resolution at the zero-coma wavelength. Figure 14.2b shows a similar pair at $\lambda = 5250$ Å. The 1000 plotted points are of random rays from the same launch point at $z = 2.56$; $x = 0$ and are of U_f

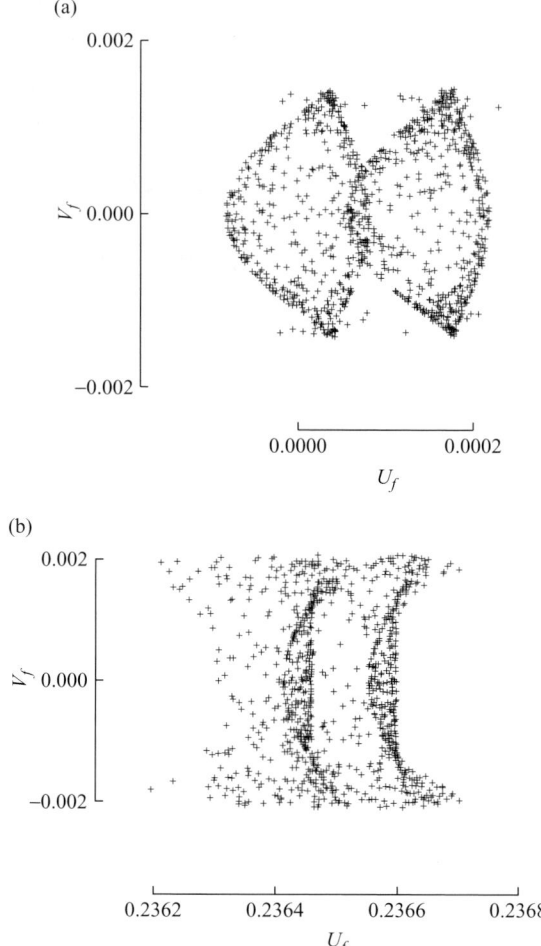

Figure 14.2 Spot diagrams of images in this spectrograph at (a) 5000 Å and 5000.15 Å at the centre of the field and (b) 5250 Å and 5250.15 Å at the long-wavelength margin. The astigmatism increases at the margin but the spectral resolution is not significantly compromised. Notice that the units are radians, the inclination to the optic axis of the outgoing rays, and that the vertical and horizontal scales are different. These figures must be multiplied by the focal length of the CCD camera lens to give the actual sizes of the spots on the CCD surface. With a 50 mm camera lens for example, the widths would be ∼7 µm and the length at the field centre would be ∼150 µm.

and V_f, the angles[4] (in radians) which they make with the optic axis. These must be multiplied by 50 to show the spot positions on the CCD. An angle of 0.0002 radian corresponds to 10 µm at the CCD surface.

---

[4] The *sines* of the angles in fact, but clearly the difference is negligible.

14.6 Requirements of a ray-tracing program

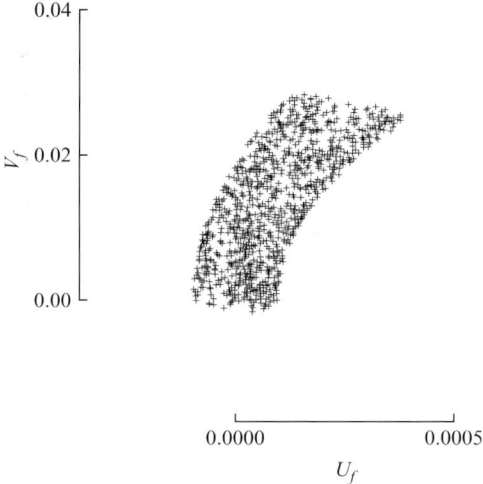

Figure 14.3 This is the spot diagram of rays from the top half of a straight entry slit 20 mm long. The horizontal scale is magnified to 100 times the vertical. Line curvature is now dominant and it may be necessary, before co-adding rows of pixels, to shift (in the computer) the upper and lower rows sideways in order to 'straighten' the image and give the best spectral resolution. Care is needed here because the line curvature varies with wavelength. There is room in this data-reduction process for considerable sophisticated software writing, since each row of pixels is, in effect, a separate spectrum with a slightly different dispersion.

Figure 14.3 shows the spot diagram for $\lambda = 5000$ Å when a finite (10 mm) entry-slit length is included. Nine rays have been launched in two annular bundles from $x = 0$ mm and from $x = 10$ mm, both directed to arrive at the same places on the grating. A thousand rays are then launched randomly from all points with $x = 0$, $z = 2.56$ and y between 0 and 10, to arrive at the same points on the grating.

There is little change in the magnitude of the aberrations but the line curvature described in Eq. (8.8) is apparent. As in the other spot diagrams, notice the different vertical and horizontal scales, which here show that with a 50 mm focal length camera lens the line images are 3 mm or 120 pixels long but that there is still less than 1 pixel width of curvature.

The final layout of the instrument was the result of many hours of investigation, and the position in particular of the M2 mirror is the position which determines the flat field at the CCD. It is some distance from the position demanded by Eq. (4.12) but takes account of the field curvature added by the focal reducer. In the end the resolving power of $\sim 30\,000$ achieved was better than the CCD resolution, and a CCD with a 10 μm pixel width would have been useable.

With a sufficiently bright source, the CCD camera could have been replaced by a traditional 35 mm camera with fine grain film.

15
Mechanical design and construction

Mechanical motions, automatic adjustments and electro-mechanical monitoring and position sensors are so diverse, and the range of transducers and electric motors for remote focusing, wavelength shifting and measuring so vast that there is no point here in listing them or criticising their relative merits. So much depends on the specification of the instrument and the capability of the available workshops that it must be left to the builder to decide how the instrument shall be controlled. The simplest possible mechanical construction is shown here – handles to turn to adjust grating turntables, capstan-bar mirror adjustments and so on, and the improvement of these by other means is clearly up to the designer and the size of his purse. This chapter therefore is confined to those matters which are peculiar to spectrograph construction.

15.1 The optical layout

The positions of the optical components will have been decided by the final version of the design coming from the ray-tracing program. The position tolerance for the mirrors is ± 0.5 mm on both x- and z-axes, the small errors being taken up when focusing.

It remains to dress them in appropriate mountings and assemble the mountings either on to an optical table or into a tube or a space-frame.

The art of constructing a scientific instrument is different from that of engineering mechanisms. The loadings are generally lighter and the precision as high as that of a machine tool or an internal combustion engine, and different criteria apply. Kinematic mounting, for example, may involve high point-loading with the consequent risk of fretting corrosion but yields a precision of fractions of a micrometre where necessary. It is suitable for optical instruments where there is no sliding motion between parts.

15.1 The optical layout

15.1.1 Kinematic mounting

Briefly, the principle is that to fix the relative positions of two solid objects, six restraints are required, three of translation and three of rotation. The common example is that of a surveyor's cube-corner reflector which is to stand on a fixed table. The cube-corner mounting has three steel balls press-fitted into holes in its base to act as 'feet' and the table has three radial grooves, each of which fixes a ball by making contact with it in two places. The device can be removed and replaced with a precision limited only by any particles of dust which may have landed in a groove at a point of contact; in other words, to a fraction of a wavelength of light. Occasionally a hole-slot-and-plane arrangement is used instead to give the six constraints, although the trihedral hole required for this is more difficult to make.

The penalty for this precision is the high point-loading at the contacts between the surfaces, and the method fails if there is a weight or contact force sufficient to distort the material significantly. It also fails when heavily loaded sliding contacts are involved. Dust in the air consists largely of micrometre-sized crystals of SiO_2 and these embed themselves in the softer of two surfaces and abrade the harder. This is a classical example of the apparent anomaly that if two surfaces of unequal hardness rub together it is the harder one which wears away. If the two surfaces are of the same material, the frictional heat generated at the point of contact may be enough to weld them together. It that case more robust methods of support are used and a ball-race, for example, will provide adequate precision with rolling friction for a grating spectrograph.

There is little in the literature on kinematic mounting and Braddick's classic work was published over 50 years ago.[1] The principles however are simple and still leave much room for inventive design.

For any optical instrument the *order* of design is important. In a Čzerny–Turner spectrograph for example the largest item is likely to be the M2 mirror, and the mounting for this will determine the height of the optic axis above the baseplate, or the internal diameter of the tube, or the datum surface if a space-frame is to be used. From this the designs of the other mirror mounts, the slit and detector holders and the grating turntable will follow.

15.1.2 The main frame

There are broadly four ways of constructing a spectrograph as illustrated in Fig. 15.1.

(1) A light alloy casting, forming an optical bench or table with raised pads at appropriate places to be milled flat to hold the optical components.

---

[1] H. J. J. Braddick, *The Physics of Experimental Method* (London: Chapman & Hall, 1954).

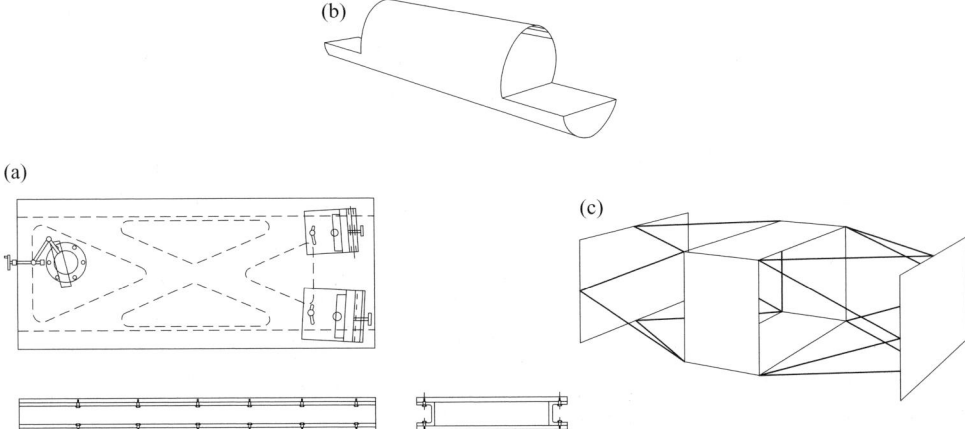

Figure 15.1 Various methods of construction for spectrographs. For many purposes the plain optical table (a) is adequate but may prove too heavy for instruments much greater than 1 m focal length. The cover, which must of course be removable, may be of sheet metal, wood or GRP, depending on the skills available. The tube (b), which is its own shroud, is appropriate to large instruments of 3 m focal length and is also very convenient for small, portable spectrographs of ∼30 cm focal length. The Serrurier truss (c) allows maintenance of the optical alignment when the direction of 'g' is changing as on a large astronomical telescope.

(2) An optical table of sheet aluminium with holes bored and threaded at appropriate places to hold the optical components. It can be a box girder and, if very large and heavy, may be self-standing, resting its Airy points on two pillars with the optical table vertical for greater stiffness. The optical components are then cantilevered out sideways on their holders. (This, incidentally, takes up much less space in a crowded laboratory.)

The covering shroud may be of metal, wood or GRP depending on the skills available.

(3) An open-frame lattice to which the optical component-holders are bolted, with a wooden,[2] GRP or sheet-aluminium shroud. If the instrument is to be attached to an astronomical telescope the Serrurier truss type of space-frame (Fig. 15.1c) is worth considering.

(4) A tube, drawn if of small (<20 cm) diameter, but probably rolled and welded if larger, with shrouded tables, plates or boxes at the ends to hold the optical components.

The choice depends on the size, the application and the quantity to be constructed. Access is always important for focus adjustment and cleaning, and large, light-tight, hinged doors at appropriate places should be part of the design, as are access holes for screwdrivers, spanners and capstan bars for fine adjustment *while the instrument is operating.*

---

[2] The author once used this type of construction with the optical shroud doubling as the packing case into which everything was packed for transport to a remote observing site.

The optical table is probably the simplest method of construction for spectrographs up to 1 m focal length. The mounting positions on the table are set by the optical design and from this the screw-hole positions follow. Threaded holes are to be preferred to through-bolting for ease of access and assembly.

Space-frame construction may be considered for large instruments, with 2–3 m focal length mirrors. This is because optical tables with adequate stiffness become unacceptably heavy and unwieldy and one must consider the feasibility of moving them from room to room and from building to building. Space-frames are more transportable although equally bulky, and the dimensions of corridors and goods-lifts[3] which may be needed to move them about must be measured. Spectrographs with focal lengths much greater than 1 m are not generally transportable. If they are to have carrying handles, then the limit is about 0.5 m.

When designing a space-frame, the criterion is, again, stiffness. Because of interference with the optical path, internal cross-bracing is scarcely possible but Warren-girder construction of welded 25 mm square-section steel tubing with solid 12.5 mm light alloy end-plates will answer the purpose. Mirror mounts similarly will act as cross-braces if designed into the frame. There is much to be said for a vertical optical table if a Čzerny–Turner mounting is specified. The space-frame is comparatively stiffer and both sides of the optical table are accessible in the event that electronic equipment is situated there. If this arrangement is chosen, the input side should be below the detector, so that the grating tilt, facing the input side, is downwards, reducing the amount of dust falling on the grating surface.

The principle of the Serrurier truss is that the frame is allowed to sag but in such a way that the ends remain parallel to each other and the optic axis is not compromised. Such an arrangement is worth considering if there is a weight constraint and the spectrograph is to be mounted on an astronomical telescope where the direction of gravity is varying with respect to the instrument during the observation. Elaborate civil-engineering design is not called for and the rule is: if in doubt make it stiffer.

A large laboratory- or Coudé-focus instrument, by contrast, is not expected to move and once it is set in position the optical alignment can be completed in the knowledge that it will remain aligned until the desired measurements are made.

15.1.3 Mirror and lens holders

The construction of these is governed by the need to avoid any sort of pressure or loading which might cause distortion of the optical surface, especially as the result of a change of temperature. Small mirrors, of diameter less than ~50 mm can be glued to a backing plate with an epoxy resin, provided that a single small area of contact

---

[3] Freight-elevators if you are an American reader!

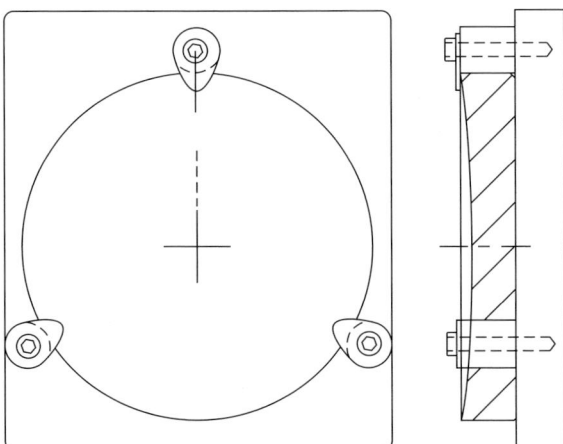

Figure 15.2 The mirror holder. For mirrors of intermediate size, in the region of 150–400 mm diameter, a simple semi-kinematic mounting is adequate. The rim rests on two of three cylindrical bearings covered by thin polythene sheaths, and is held by retainers of beryllium–copper screwed to the ends of the bearings, exerting minimum pressure on the mirror.

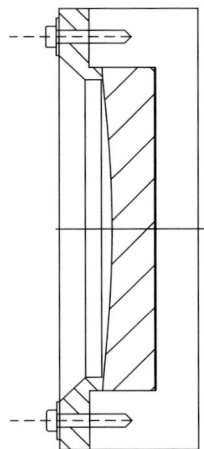

Figure 15.3 The mirror cell. This is suitable for smaller mirrors of a few centimetres diameter. The mirror rests on a polythene pad and is retained by a counter-cell held in place by screws (*not* threaded as lens counter-cells sometimes are).

(~5 mm diameter) is sufficient to carry the weight. Multiple attachment points must be avoided because of the bending which may result from differential expansion. Mirrors and lenses larger than this require either a cell, or quasi-kinematic support. Figure 15.2 illustrates a suitable mounting for mirrors of 150–400 mm diameter. Mirrors smaller than this may be contained in a mirror cell such as that in Fig. 15.3,

15.1 The optical layout

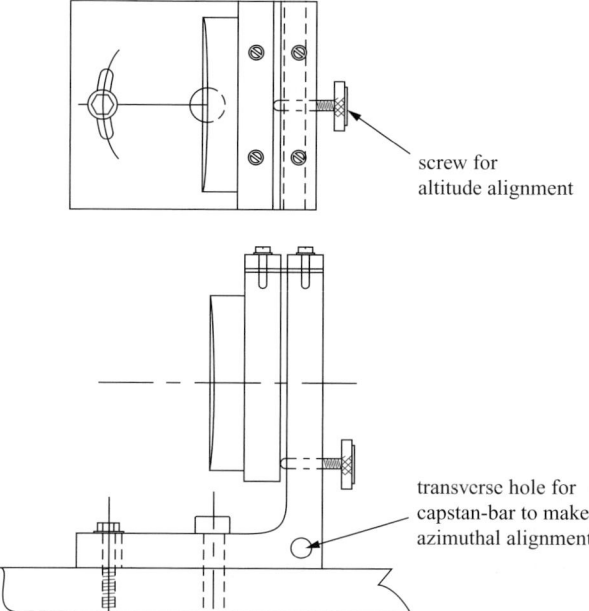

Figure 15.4 The mirror mount. The principle is that the mirror can be rotated in azimuth and altitude almost exactly about its vertex and, once adjusted satisfactorily, is fixed in place by a single bolt. The same mounting is suitable for gratings when no rotation is needed to change the wavelength range.

where they are backed by a thin polythene film and held in place by a counter-cell fixed by screws *which are not over-tightened.*

The thin polythene film is used to avoid high local pressure, which may damage brittle materials like glass. The accuracy with which a mirror is positioned is not high by engineering standards, and an error in centration[4] of 0.1 mm is generally acceptable when focal lengths are of the order of 1 m.

15.1.4 Mirror adjustments

The mirrors require adjustment and an altazimuth set of axes is convenient and practicable. In Fig. 15.4, the mirror holder may be attached to an L-bracket by means of a horizontal beryllium–copper strip-spring clamped in place with a keeper, and the L-bracket is rotated about a vertical axis which passes through the mirror vertex. To ensure this is a simple matter of design and the rotation is secured by a journal bearing – a simple drilled hole in the baseplate with a brass dowel passing through a similar hole in the L-bracket. This arrangement works well for mirrors

---

[4] Centration here means the accuracy with which the optic axis of a lens or the vertex of a paraboloidal mirror coincides with the centre of the circular rim.

up to 400 mm diameter. The L-bracket is clamped in place with a bolt fitting a threaded hole in the baseplate through an arc in the L-bracket base. A washer at the bolt head is vital here to retain the accurate positioning of the azimuth as the bolt is tightened.

The optical alignments are then carried out as follows: the tilt about the horizontal axis is governed by a brass screw with a 1 mm pitch bearing against the rear of the mirror cell and held in contact either by the force of gravity or if necessary by a retaining spring. The adjustment about the vertical axis is by a silver-steel capstan bar inserted into a hole at the elbow of the L-bracket. The capstan bar should be long enough to allow the positioning of the mirror to $3'$ of arc – a comfortable tolerance in practice. Multiple attachment of the L-bracket to the baseplate is not necessary, provided of course that the single bolt is big enough.

If the instrument is subject to shaking or vibration a more solid fixing is needed. In this case three pairs of push-pull screws will hold plates against plates essentially as a solid, and the holding of the mirror in its cell becomes a more difficult task when allowance is made for differential thermal expansion and the brittleness of glass and fused quartz. An oversized cell may be used with a gap around the mirror edge approximately three times the thickness of the beryllium–copper leaf-springs which are inserted to keep the mirror centration exact. A front-plate, also of beryllium–copper, with a 3 mm overlap at the edge, will do no more than obscure the turn-down at the edge of the mirror. These extra precautions are only needed if a very large temperature range – between midday and midnight in the desert for example – is expected.

For small mirrors, of less than 50 mm diameter, the vertical axis is sometimes carried on a beryllium–copper strip-spring, but this foregoes the opportunity to have the axis of rotation through the vertex, and requires a holding spring to locate the azimuth adjustment.

15.1.5 Grating holders

Much depends on the portability of the finished spectrograph. In the case of a large, immobile instrument with a horizontal optical table, the grating may be held free as in Fig. 15.5, resting on a lower pad of stiff plastic such as polythene, and constrained by a light alloy bar which rests on an upper polythene pad and in turn is held in place by capstan-head screws. The rear face of the grating may be insulated from its mount by a thin sheet of polythene.

If a cell is required it may be fabricated, or milled from a single block of light alloy, with relieving holes drilled at the corners as in Fig. 15.6. Again space should be left for insulating polythene sheets and if necessary PTFE clamping screws will hold it immobile. Gratings are generally formed on quartz blanks and the low

15.1 The optical layout 157

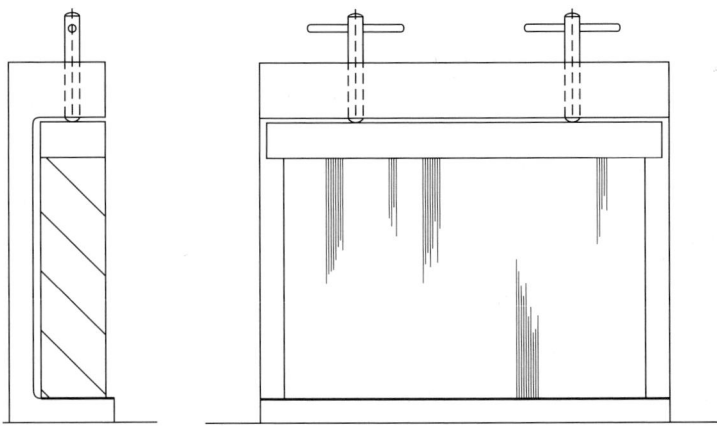

Figure 15.5 The grating cell. The grating can stand on a shelf at the appropriate height above the table, and is retained in place by a bar exerting minimum downward pressure through two capstan screws.

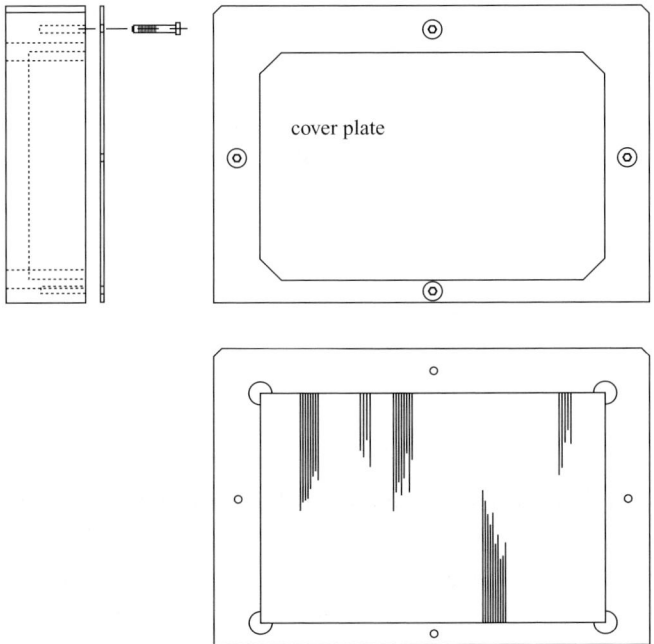

Figure 15.6 An alternative grating cell, in which the grating is fully enclosed and retained by a face-plate of light alloy with an appropriate mask-shape as in Fig. 8.5.

thermal expansion of these must be borne in mind. The grating may be held in its cell by a face-plate which restrains the corners, flexible enough that when the screws are fastened it places only the necessary pressure on the grating, sufficient to maintain it against temperature changes and to bear its weight when it is possibly face-down at the Cassegrain focus of a telescope.

Optical adjustments are probably needed only in the initial alignment process, in which case no elaborate rotation about the grating normal is called for. Rotating the grating in its own plane to align the rulings can be done with copper or bronze shims under the lower edge.[5] A metal shim, a sliver 25 µm thick, under one end of a 150 mm wide grating will tilt the rulings by 0.5' of arc.

15.1.6 The grating turntable

In a high-resolution spectrograph it would be unusual if the whole of the spectral range were available to the detector without rotating the grating. If the adjustment is infrequent the grating cell can be mounted on a circular brass pad which rotates about a simple journal bearing in the form of a spindle lying in a hole in the optical table. It is then held in position by a clamping screw, much like the provision of the mirror holders.

If adjacent regions of the spectrum are to be photographed in rapid succession a more elaborate arrangement is needed and a proper turntable must be used. The standard method in this instance is to use two large-diameter, light-duty angular contact bearings in opposition, held in permanent contact by means of a compression spring as in Fig. 15.2. A simple worm-gear can be used for rotation, but there is an advantage in using a sine-bar, as in Fig. 15.7. The valuable property of a sine-bar motion is that the wavelength presented at a particular pixel varies linearly with the angle turned by the screw. This in turn leads to convenient calibration and re-setting. In a scanning monochromator, such a drive, usually motorised, is regarded as virtually essential. Of various possibilities in practice, the double-hinged version is to be preferred. The moving disc bearing against a ball engenders wear and fretting corrosion with consequent loss of calibration and smoothness of scan.

In the asymmetrical Čzerny–Turner mounting the grating face must make an angle $\alpha - \beta$ with the sine-bar to which it is fixed since the grating equation, Eq. (8.2), which governs the relation between captive-nut position and grating tilt angle, t, can be rewritten as

$$n\lambda/a = 2\cos(\alpha + \beta)\sin(t + \alpha - \beta). \qquad (15.1)$$

---

[5] With good modern workshop practice this will almost certainly not be necessary.

15.1 The optical layout

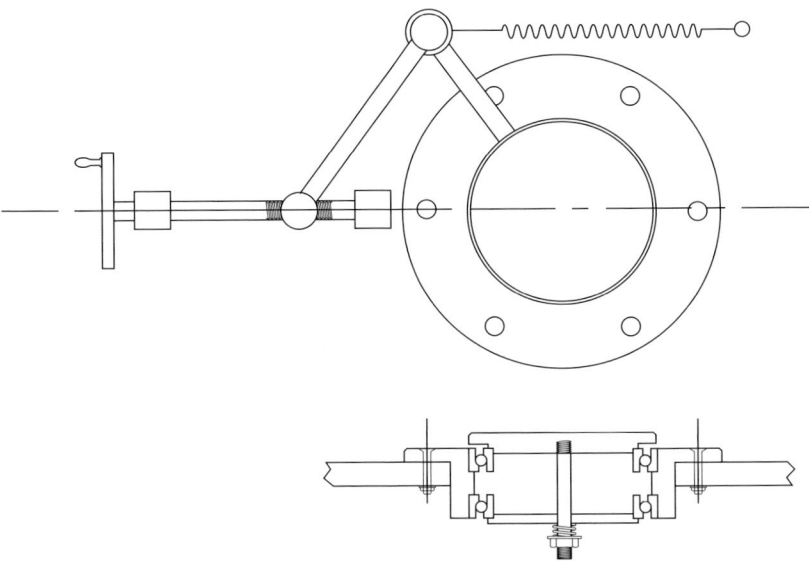

Figure 15.7 The grating turntable. In most cases it is not possible to record the entire spectrum in one exposure, and different regions are brought into view by rotating the grating. It is important to be able to do so accurately and for this a grating turntable is required. A standard form is shown here with manual adjustment which can obviously be replaced by a stepper-motor under computer control. The sine-bar link ensures that the screw moves the central wavelength of the spectrum linearly with rotation angle.

In the original Ebert mounting α and β were the same and $\alpha + \beta$ was called the ebert angle.

Whichever method is used, there is backlash in the drive screw to be considered. Although there exist sophisticated anti-backlash split-nuts and similar devices, a light pressure applied to the turntable by a spring is adequate for most purposes.

15.1.7 Entry and exit slits

Many constructors have spent time, ingenuity and energy on manufacturing double-opening slits, where both jaws move apart symmetrically. This is probably a mistake. Traditionally only one jaw moved and the other, fixed jaw was the fiducial mark for measuring deviation angle in a prism spectroscope. An emission line spectrum could then be observed with slits as wide as convenient to allow plenty of light to enter the eye, and this in turn made for easy location of the fixed edge with a pair of cross-hairs at 45° to the line of the slit edge.

There is still no particular virtue in the double-opening slit. If wavelengths are to be measured, the fixed edge is still the fiducial mark even on a photographic plate,

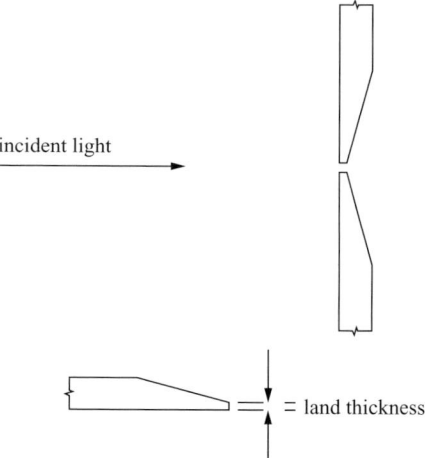

Figure 15.8 A section through the slit jaws, illustrating the lands of the jaws. Light should always be incident on the plane faces of the jaws.

and unless there are close doublets to be resolved, there is no virtue in a narrow slit width when observing emission spectra.

Slits should always be mounted with the light incident on the plane, unchamfered face as in Fig. 15.8. The purpose of the chamfer is to permit a slit width as small as 10 µm with the thickness still less than the width. In other words, slit edges must be knife-sharp.

Much ingenuity has been expended on slit-adjusting mechanisms. The simplest appears to follow the practice of R. V. Jones's parallel motion, a device much used in instrumental practice. It is illustrated in Fig. 15.9. A ball-ended micrometer pushes its ball against an anvil such that the point of contact is midway between the moving and fixed bars. The width of the slit is read *directly* off the micrometer which can have a large diameter drum for easy reading in dim light. The arrangement is such that the slits cannot be damaged by over-tightening of the micrometer and a preventer screw holds the moving jaw at the point of closure to avoid damage to the slit lands.

Slit jaws themselves should be replaceable. They are normally made of hardened and ground tool-steel, or 'gauge-plate' as it is sometimes called, and can be made straight enough with a grinder that has not previously been misused in the workshop. If ultimate precision is needed, then three similar edges can be ground and lapped alternately in the same fashion employed in making optical flats.

It is probably convenient to shot-blast or grit-blast the plane faces of the input-slit jaws before the final grinding of the edges, so that they act as a focusing screen for the fore-optics.

The slit assembly will normally be attached to a mounting, which itself may stand on the optical table, but which may more conveniently be fitted to the edge

15.1 The optical layout

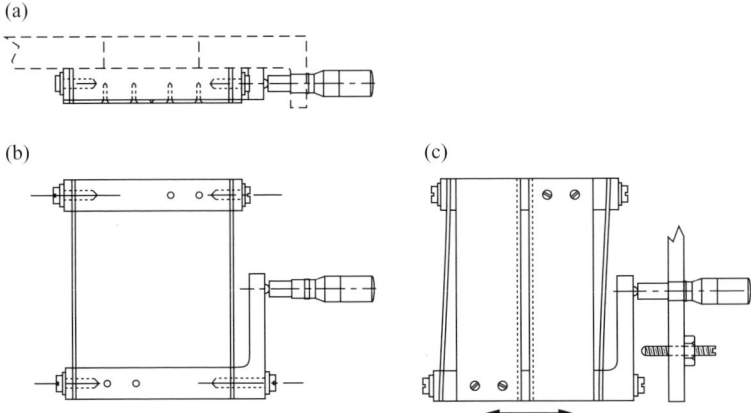

Figure 15.9 The parallel motion for opening slit jaws. One jaw is attached to the upper, fixed bar and the other to the lower L-shaped bar which moves when pushed by the ball-ended micrometer. The anvil of the micrometer must always be midway between the two bars.

of the table. Some attention must then be given to any fore-optics which may be attached to this mounting and sufficient strength must be given to enable it to bear the weight and moment. The classical approach was to mount an optical bench along the input optical axis, but this is sometimes inconvenient and such things as external baffles, colour filters and order-sorters might as well be made with threaded ends to be attached directly to the entry-slit mounting, much as they are on camera lenses.

15.1.8 The CCD camera attachment

The LN_2-cooled CCD camera may be a common-user component which must be removed occasionally from one instrument to another, or from a single instrument for focus adjustment and cleaning. In any case a quick-release mechanism is desirable. One possibility is the hook-on-shelf attachment shown in Fig. 15.10 which allows adequate precision, which avoids the need for extensive screw-fittings and which interfaces easily to different spectrographs with a simple common interface. If the attachment is to an astronomical telescope, one substantial additional bolt will secure the camera against variations in the direction of the gravitational g-vector.

15.1.9 Access

It is likely from time to time that the alignment and focus will be checked and it is convenient if this can be done while the instrument is in a well-lit laboratory. Access ports allowing a hand or a screwdriver shaft to reach in and turn an adjusting knob

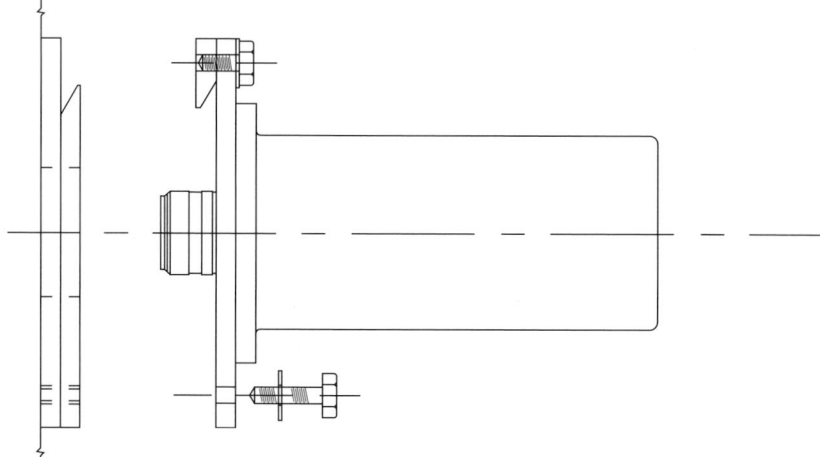

Figure 15.10 A quick-release attachment for a CCD camera. This was designed to attach a Wright Instruments LN_2-cooled camera to two different spectrographs. The essence, which is easily adaptable for other types of camera, is a pair of interlocking hooked shelves which bear the weight and allow the camera to be slid horizontally against a stop to achieve correct alignment. If necessary the binding can be made firm by a screw at the bottom. The camera can be released and removed to another spectrograph with a similar fitting in a matter of seconds.

should be incorporated at suitable places, and in particular a means of adjusting the grating table from behind the grating will probably be required. Access holes must be planned in advance before the shroud is made.

The shroud itself as well as the base should be thermally insulated to give the greatest possible temperature relaxation time. This is not so important in the normal thermostatted laboratory, but the interior of the dome of a 2-metre telescope may fall in temperature by 10–15 °C during a night's observation, with a corresponding change in length of a 1-metre spectrograph optical table of a millimetre or so unless the insulation is adequate.

15.2 Optical materials

Mirrors and grating blanks are generally made of fused silica, ground and polished to an accuracy of '$\lambda/8$ of green light'. This is to ensure that a reflected plane wave from a grating for example remains flat to $\sim\lambda/4$. Their advertised focal lengths are not generally accurate to better than $\sim 0.3\%$ and should be measured with a Foucault knife-edge test. The measured focal lengths may then be used to revise computed values in the ray-tracing program.

Mirrors are made more cheaply from low-expansion borosilicate glasses and more expensively from vitreous ceramics with a zero thermal expansion coefficient over a limited temperature range. Such materials are also useful for interferometer construction.

Transparent materials include various crown and flint glasses for the visible and near ultra-violet, fused silica for the ultra-violet, calcium fluoride and lithium fluoride for the extreme ultra-violet as far as 1200 Å. Below this wavelength nothing is transparent until the soft X-ray region, and gas retention for example must rely on thin films of collodion or some similar plastic material which can be formed into films 1 μm thick. Such films tend to be porous and a double-layer film is to be recommended. They will withstand a pressure difference of 5 mm Hg over an aperture of ∼3 mm diameter. Since gases are as opaque as other matter, this is an adequate pressure for absorption spectrometry in the VUV.

15.3 Transparent optical materials

Glasses are available with different optical and physical properties and can be chosen by refractive index, which is between 1.44 and 1.95, by dispersion or Abbé v-value, by weathering properties, which makes them sometimes unsuitable for the outer component of a lens, and not least by cost, with extreme examples of high-index lanthanum crowns being beyond the purse of most scientists. Glasses are not generally transparent below 3000 Å.

For most general purposes Schott BK7, a borosilicate crown appears to be the glass of choice among makers of singlet lenses. The Schott catalogue of optical glass is the bible on the subject.

In the infra-red a whole gamut of transparent materials exists with refractive indices up to 2.5 or higher. These include polycrystalline pressings and sophisticated plastics materials. Among these are sulphones which have Abbé v-values less than 20, implying enormous dispersion and possible use in prism spectrographs. The Eastman Kodak company's products will fill most transparency requirements in the infra-red. For reflection, a gold coating has the highest reflection coefficient.

In general a study of the properties of transparent materials is not necessary for the construction of a spectrograph as nearly always a commercial product – especially a photographic lens – will be found for most purposes. The question chiefly of interest to the spectrograph designer is the refractive index, dispersion, transparency and cost.

15.4 Reflectors

The choice between a spherical mirror and a paraboloid depends to some extent on the optical requirements, but at focal ratios above F/10 and focal lengths below 1000 mm the difference is not significant. The surface error of a sphere, that is its

164   *Mechanical design and construction*

departure from a paraboloid, is $\Delta x = y^4/64 f^3$, where y is the semi-aperture and f the focal length. Alternatively, the maximum tolerable aperture, D (millimetres), where parabolising makes no difference, is given by $D = 0.064 F^3$ where F is the focal ratio.

Mirrors in optical instruments generally have a diameter/thickness ratio of about 6:1 to give the necessary elastic stability. Diameters up to 600 mm are solid but when weight becomes a problem, then a honeycomb structure may be indicated. Interferometer plates, especially for the visible or near UV may be proportionately thicker, with a diameter/thickness ratio of 4:1. They will invariably be made of fused silica.

15.4.1 Cleaning optical materials

Mirrors will probably arrive with the reflection coating in place and this will have been overcoated with silicon monoxide or some similar protective layer. The surface is fairly durable and can be cleaned, but prevention being better than cure, they should be protected when not in use with covers to exclude dust, insects and oily vapours which may condense.

The surface of an aluminised mirror will be hard enough that, in the words of one eminent astronomical telescope mirror-maker, 'it can be washed with anything you would use to wash your face.' Nevertheless, most people will be more cautious and will clean their optical sufaces with de-ionised water,[6] analytical reagent grade iso-propyl alcohol and a soft cotton-wool or tissue swab which has been removed from a just-opened packet. The point is not trivial: aerosol particles of silica a few wavelengths in diameter, which are universal at sea level, falling on to a clean polished surface, are wonderfully abrasive. Cleaning cloths should always be kept folded, should be washed frequently and dried in a closed desiccator, and only the inside of the folded cloth should be used. The blower brushes commonly used with camera lenses may be of inadequate size for 200 mm diameter mirrors. Compressed air may not be free of oil droplets.

Uncoated lens surfaces may be cleaned with the aforementioned recently unfolded cloth, and if there is a natural bloom to be removed, jeweller's rouge can be applied before rubbing gently. Surfaces with anti-reflection coatings should not be treated thus and cleaning is confined to the usual solvents softly applied.

15.4.2 Diffraction gratings

It goes without saying that the grating surface must not be touched. However, grating surfaces do become contaminated eventually and they can be cleaned. Although

---

[6] This includes exhaled breath, a perfectly adequate cleansing fluid.

no solid may make contact with the surface, suitable liquids will remove dust particles and lipid condensation. Analytical quality chemical solvents such as iso-propyl alcohol and diethyl ether, used in well-ventilated, warm, dry surroundings, poured gently over a grating held at 45° to the horizontal in a photographer's developer tray will make a remarkable restoration to a contaminated grating: but solvent evaporation, particularly of ether, will cool the surface and lower the local dewpoint so that water vapour may condense on to the grating and leave a faint stain.[7] Therefore finish the cleaning with iso-propyl alcohol.

15.5 Metals for construction

The days when scientific instruments were made of brass are unhappily long past. Light alloys of one sort or another are the materials of choice both for laboratory and field instruments, with various sophisticated lightweight honeycomb materials for space-borne instrumentation, with metal inserts at strategic places to support the items which require firm anchorage.

Thanks to computer-controlled machine tools, fabricated light alloy construction is not the time-consuming bugbear it once was and parts once fitted and bolted together are now machined integrally. Nevertheless, light alloy castings for the main optical table are worth considering, especially for instance if several examples are to be made for a teaching laboratory; and there is then a considerable saving of workshop time because of the comparatively small number of machining operations to be done.[8] Set against this is the need to 'mature' the casting, which may be liable to gradual 'creep' and distortion. Traditionally, iron castings were laid on a flat roof to 'weather' for a year, but this is probably excessive with modern light alloys and casting techniques. The largest casting that can be accepted by the available workshop machine-tools is also a factor, especially for spectrographs of more than a metre focal length. Turning is an unlikely process for large castings but the milling of several separate surfaces to the same plane requires an adequately large milling machine.

Bolted sheet aluminium alloy is a convenient method of construction for small to medium-sized instruments. Welding is not recommended where strength is essential and epoxy-gluing may be too permanent when possible later modifications are required. The weight advantage of magnesium and its alloys is completely offset by its instability and brittleness and magnesium is generally worth considering only if the weight-saving requirements are extreme. Brass and gunmetal are expensive but useful for journal bearings.

---

[7] The ideal environment for cleaning a grating is a dust-free room in the middle of the Sahara Desert.
[8] In the vicinity of rust-belt cities, search the scrapyards for superannuated lathe-beds and milling-machine tables, which make superbly stable optical tables: but be sure your workshop can handle the necessary modifications.

The bending tolerance for the optical table is ~0.5 arcminute, rather than the 0.01 arcsecond required in an uncompensated interferometer. As a guide, adequate stiffness for a 1-metre spectrograph is achieved by using two parallel light alloy plates 12.5 mm thick, 1.2 m × 0.4 m in area, connected by two parallel U-section girders 100 mm × 60 mm of 10 mm thickness. The weight can be relieved by removing those sections of the lower plate which do not affect the torsional stiffness.

The whole structure, in accordance with kinematic principles, should be supported on three feet, placed at the Airy points which, for an unencumbered beam of length L are at a distance $L(\sqrt{3} - 1)/2\sqrt{3} = 0.211L$ from the ends. This balances out sag and hump and ensures that the ends are horizontal. Bear in mind that the weight of the optical table is considerably greater than the combined weight of the optical components which will stand on it.

The qualities of alloys, especially aluminium alloys, vary according to composition, and some are more amenable to turning and milling than others. Materials to avoid are (1) pure aluminium, (2) stainless steel except in finished form such as nuts and bolts, (3) mild steel, which corrodes easily and (4) pure copper, which is ductile and difficult to shape with machine tools.

Beryllium–copper[9] in thin sheet form is superlative for making leaf-springs, but after forming it must be 'tempered' (precipitation hardened in fact) by boiling it in benzyl benzoate under reflux for 90 minutes. Thicknesses between 0.5 mm and 1.0 mm are suitable for most laboratory instrument spring-hinges. All the forming, cutting and drilling must be complete before the tempering.

Plastics such as polythene and PTFE (poly-tetra-fluoro-ethylene) have appropriate properties for insulation, padding etc. PTFE is machinable, durable and soft, but has a large thermal expansion coefficient. It is also available as screwed-thread, nuts, bolts and washers. Washers should always be used *on both sides* with nuts and bolts and with bolts fitted into threaded holes. Bolts in threaded holes are always to be preferred to nut-and-bolt fastening unless there is a risk of overtightening and consequent stripping of the thread.

15.5.1 Journal bearings

Bearings for light loads, whether journal or sliding, may be simple contact sliding faces. They should *always* be of different metals: light alloy and brass for example or silver-steel and gunmetal. Sliding faces of light alloy, in particular, will weld together sooner rather than later and must be avoided. Any sort of grease or oil is anathema in a spectrograph and all lubrication should be dry, either by graphite or

---

[9] Copper alloyed with 2% beryllium, much to be preferred to phosphor-bronze which can only be hardened by cold working.

by a colloidal suspension of graphite such as that sold under the name of 'Aquadag'. For heavy loads or for precision rotation, ball-races should be used[10] and circulating ball-race bearings will support translational motion. They are not lubricated and must be protected from dust.

15.6 Other materials

An American friend and colleague, a virtuoso at interferometer construction, now sadly deceased, occasionally made his prototype Michelson Fourier spectrometers out of wood. This is not recommended for ordinary mortals. He also believed in rapid scanning to give an interferogram every few milliseconds, co-added until the signal/noise ratio was satisfactory. This, by contrast, has its merits.

Wood nevertheless is an acceptable material for the shroud – the light-tight cover of a spectrograph. It is light in weight for easy removal, is easily shaped, with cut-outs for access and for sliding over necessary metal protrusions. For field instruments, marine ply has excellent strength/weight ratio and can be made weatherproof and durable. The alternatives, for those skilled in the art, are either GRP (glass reinforced plastic), the material from which boats are constructed, or frameworks clad with sheet-metal.

15.6.1 Protective coatings and paint

Zinc chromate is the accepted protective coating for light alloys but is probably unnecessary unless the finished instrument is to stand in the open air. Even then it must be overcoated with a suitable weatherproof paint. For laboratory work light alloys can be anodised and dyed, and black-anodising of a previously shot-blasted or abraded surface is a convenient and durable way of treating the interior. Otherwise matt black paints such as 'optical black' can be applied with an appropriate undercoating according to the manufacturer's instructions.[11] Beware of the peeling of paint from unprepared metal surfaces. Exterior fittings and controls are traditionally painted white for easy visibility in low-light conditions and it is worth making different hand-controls of different shapes so that they can be recognised in the dark by 'feel'. It sometimes happens that it is necessary to work in complete darkness and pilot-lights, LEDs and such should have covers which can be closed. The extreme sensitivity of CCD spectrographs to faint light must not be disregarded.

---

[10] Except in interference spectrometers where kinematic hinges should be used.
[11] There is no such thing as 'black' paint. There are various shades of dark grey, or sometimes dark blue or dark green, and a low albedo is all one can hope for. Some 'black' paints are anything but black in the infra-red and the *shape* of a baffle is more important than its colour.

16
Calibration

16.1 Sensitivity calibration

There are so many variables in the optical path of a spectrograph that sensitivity calibration is necessarily totally empirical. A black body at a known temperature is the only feasible method of absolute calibration, and even then one must be sure that there has been a total suppression in the spectrograph of other orders and of scattered light. Reliable *absolute* calibration is such an onerous task that one must question the need for it in most circumstances. For almost all practical purposes a comparison with a standard source is sufficient. In the visible, a reasonable approximation – no more – can be obtained by making a spectrogram of a surface coated with freshly deposited magnesium oxide from burning magnesium ribbon which will scatter the light of the midday sun on a clear day. In the tropics and at moderate temperate latitudes, the midday sun is an approximation to a black body at 6000 K. Ordinary white card will not do as a scatterer, neither will a painted white matt surface because both are likely to be fluorescent under the UV solar radiation.

The resulting spectrum, after correction for the CCD pixel-to-pixel variations of sensitivity, may not look much like a textbook black body spectrum and for good reason. Firstly there are the basic variations with wavelength of the CCD overall sensitivity. Added to these are the reflection coefficients of the various mirrors and above all the variations with wavelength and polarisation of the grating rulings and the change of these with angle of diffraction. (Spectro-polarimetry with a reflection grating is a lost cause for this reason.) Nevertheless, since the radiation intensity is calculable from Eq. (3.1), the spectral sensitivity of the spectrograph can be estimated.

16.2 Wavelength calibration

The most obvious and most useful way of making a wavelength calibration curve or table is to select ten or twelve known wavelengths covering the spectral range required, as evenly spaced as circumstances permit, and to tabulate them against pixel number, then to draw or compute a cubic spline[1] to fit them. In general, with a grating spectrograph, the calibration curve of pixel number vs. wavelength will be close to a straight line and a table of interpolated wavelengths vs. pixel number will not deviate by more than ~0.1 Å from the true values. Single sharp emission lines will usually occupy two adjacent pixels and *if the line is symmetrical*, its wavelength can be interpolated using the ratio of the two pixel counts. Causes of line asymmetry can be physical (e.g. pressure broadening) as well as optical shortcomings, chiefly the result of coma, in the spectrograph.

There are many cubic-spline-computing programs in various computer languages to be found as public domain software, ready for loading, compiling and using.

The output from a cubic spline program should be available as a table of wavelength against pixel number as well as a graph.[2] The spectrogram, binned down to a one-dimensional table of intensity vs. pixel number by the camera software, together with the spline program output, can then be loaded into a spreadsheet program and thence plotted as a perfectly linear $x-y$ graph of intensity against wavelength. Various arithmetic functions can be applied, such as subtraction or division of successive spectrograms, to measure such things as time variation, optical filter profiles and absorption coefficients.

16.2.1 Infra-red calibration

There is a relative poverty of suitable sharp emission lines for wavelength calibration in the infra-red and it may be necessary to resort to the method of Edser–Butler fringes.[3] These are most suitably generated in a Fabry–Perot étalon with metal coatings and quartz Fabry–Perot plates. These will give an adequate finesse ($F \simeq 10-12$) to allow wavenumber measurements to be made. Accuracy depends on knowing precisely the gap width between the Fabry–Perot plates, and this is measured if necessary by the (tedious) method of exact fractions with known calibration wavelengths in the visible.[4] Bear in mind that the Edser–Butler fringes are

---

[1] Cubic splines are what you get when you plank the ribs of a ship with single long strakes of wood. Drawing-pins fixed on a drawing board, with a length of fine piano wire threaded so as to touch each pin, give the same effect.
[2] The graph is a useful weapon against misplaced entries in the data, which show up instantly in loops and whorls in the plotted graph.
[3] These are the fringes obtained when λ, in Eq. (4.1), is the independent variable in ϕ and I is plotted as a function of ϕ in Eq. (4.2).
[4] See for example, R. W. Ditchburn, *Light* (New York: Dover, 1991), Chapter 9.

at equal intervals of wave*number*, so that it is important that the gap be an adequate multiple of the longest wavelength to be recorded by the spectrograph under calibration.

16.3 Small spectral shifts and radial-velocity measurement

A well-focused CCD grating spectrograph has a remarkable ability to detect small shifts of wavelength, provided one is confident of its thermal stability. Thermal insulation is important: one may alter the calibration just by standing too close, allowing body-heat to warm one side of the optical bench, expanding it and so distorting the optical path. One 25 μm pixel subtends $\sim 9''$ of arc in a 1-metre spectrograph. The ultimate resolution at 5000 Å typically is 0.015 Å, subtending $\sim 1''$ of arc, and a shift of illumination of 5% from one pixel to its neighbour implies a wavelength shift of $0.5''$ of arc, i.e. 0.008 Å or 0.5 km s^{-1} in astronomical terms. This is – just – enough to reveal the morning-to-evening Doppler shifts of solar Fraunhofer lines caused by the diurnal rotation of the Earth. A laboratory lamp to furnish a calibration wavelength is essential of course.

16.4 Absorption measurements

Absorption spectroscopy is traditionally the most difficult spectroscopic technique[5] and the precautions to be taken are severe. The shape of an absorption line is modified by the instrumental profile of the spectrograph and an accurate line profile may not be available. Astronomers in particular are content often with the *equivalent width*, as this measures the amount of power removed from the continuum by the absorption, which is the usual purpose of the investigation. Elaborate precautions must still be taken to exclude scattered light, but provided this is done, the equivalent width is not affected by the instrumental profile. Equivalent width is determined as follows.

Assume the spectrum is normalised to the continuum level. If the spectrum in Fig. 16.1 is $1 - S(\lambda)$ and the instrumental profile is $I(\lambda)$,

$$\int_a^b S(\lambda)\,d\lambda = S$$

and S is the total power removed from the continuum by the absorption feature.

By definition:

$$\int_{-\infty}^{\infty} I(\lambda)\,d\lambda = 1$$

---

[5] Especially in XUV region where it is next to impossible.

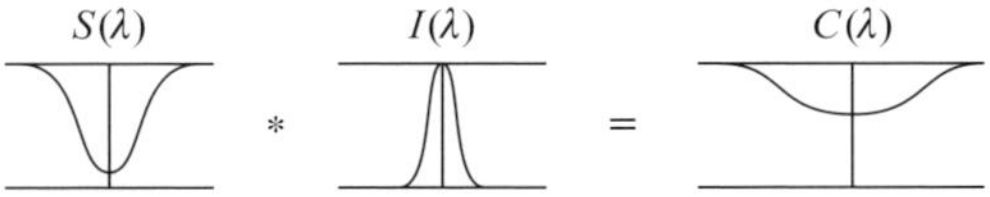

16.4 Absorption measurements

Figure 16.1 The principle of the equivalent width measurement. The true absorption line $S(\lambda)$, when convoluted with the instrumental profile $I(\lambda)$, produces a shallower, broader absorption line in the recorded spectrum. Provided there is no contribution to the recorded continuum by scattered light, the area of the absorption line is the same as that in the original, and measures the total amount of continuum radiation that has been absorbed. The equivalent width then is the width of an idealised rectangular absorption feature of full depth with the same area.

and if the limits a and b are far removed from the absorption region we can replace them as usual by $\pm\infty$.

The output from the spectrograph, which is the convolution of the two, is

$$\int_{-\infty}^{\infty} S(\lambda) I(\lambda' - \lambda)\, d\lambda = C(\lambda')$$

then

$$\int_{-\infty}^{\infty} C(\lambda')\, d\lambda' = S$$

because with $\lambda' - \lambda = y$, $d\lambda' = dy$ so that

$$\int_{-\infty}^{\infty}\int_{-\infty}^{\infty} S(\lambda) I(\lambda' - \lambda)\, d\lambda = \int_{-\infty}^{\infty}\int_{-\infty}^{\infty} S(\lambda) I(y)\, d\lambda\, dy,$$

which separates into

$$\int_{-\infty}^{\infty} S(\lambda)\, d\lambda \int_{-\infty}^{\infty} I(y)\, dy = S.$$

The equivalent width is thus defined independently of the spectrometer instrumental profile as $\Delta\lambda$, where $I(\lambda)\Delta\lambda = S$. In other words the area still measures the amount of continuum power removed from the spectrum by an absorption line.

For those occasions where there is a need for the absorption line profile – for example where there are narrow deep cores to comparatively broad shallow lines, showing perhaps that a cool, optically thick outer layer of stellar atmosphere or an interstellar cloud has been traversed by the radiation – there is no option but to use a very high resolving power so that the slit width is narrow compared with the half-width of the absorption line (which is not always well defined).

17
The alignment of a spectrograph

The method described here is suitable for the assembly and adjustment of a Čzerny–Turner type of spectrograph. Essentially similar methods are indicated for other types and mountings.

Alignment is best carried out systematically after the instrument has been installed in its place of work or, if it is newly made, close to the workshop which made it.

17.1 The optical alignment

The optical table has a number of holes drilled at the vertices of the mirrors and the grating. Into these holes will normally fit the vertical axle about which the angular adjustments of the optical components are made. If a precision milling machine is available, similar holes can be drilled along the optic path so that alignments can be checked at any time.

An alignment tool (Fig. 17.1) should be made, preferably but not necessarily of metal, which will stand on the optical table with a base extension which will fit these holes. It should carry a vertical disc with a 1 mm hole at the height of the optic axis.[1]

Stage 1. A laser beam, adjusted to the height of the optic axis, is made to shine through the centre of the entry slit so that it is parallel to the optical table surface. 'Parallel' here means that the light from the laser would be able to pass through the hole in the alignment tool no matter where the latter is placed on the table.

Stage 2. The laser beam direction is now adjusted until it shines towards the vertex of the collimator (M1) mirror. The alignment tool, placed in the hole corresponding to the vertical axle of the mirror mount, has the central hole at the mirror vertex position. The tool can then be removed and the M1 mirror installed. The procedure ensures that the beam is shining along the optic axis.

---

[1] I am indebted to colleagues at Perkin-Elmer Ltd. for this method of optical alignment.

17.2 The focus

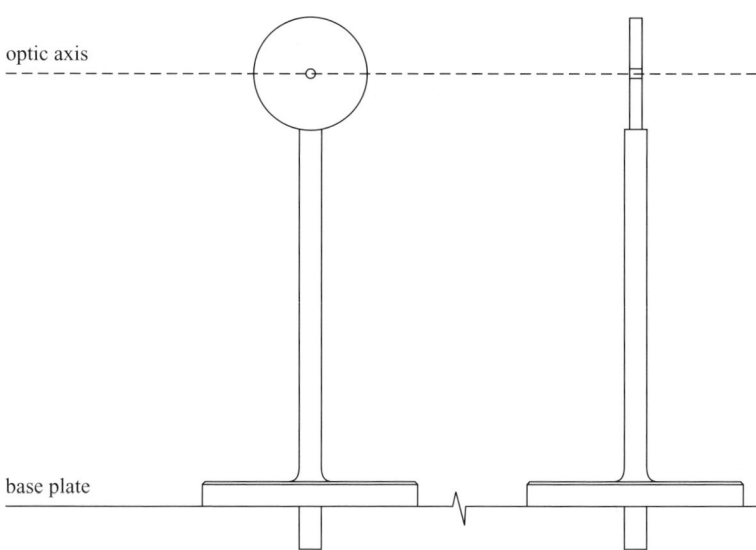

Figure 17.1 The optical alignment tool. Its use depends on having holes for the dowel drilled in the spectrograph baseplate along the optic axis, with x- and y-coordinates accurate to ~ 25 µm. When a laser beam shines through the central hole with maximum intensity, the alignment in a 1.5-metre spectrograph is good to better than $1'$ of arc. The dowel holes should be close to (<25 mm) the mirror and lens vertices and to the grating surface.

Stage 3. The alignment tool is next transferred to the grating table vertical axle hole and the M1 collimator mirror is adjusted in both axes until the reflected laser beam passes through the hole in the alignment tool. The tool is then removed and the grating with its mount is installed in its proper place.

Stage 4. The alignment tool is transferred to the vertical axle of the focusing (M2) mirror and the grating adjusted in zero order so that the light passes through the alignment tool. When this adjustment is satisfactory the grating should be rotated until the first-order reflected beam passes through the hole in the alignment tool. It is at this point that the grating may need shimming in case the rulings are not parallel to the rotation axis. The adjustment is then checked by rotating the grating so as to pass other orders through the alignment hole.

The M2 mirror and any auxiliary optics are installed, and the M2 mirror is adjusted until the laser beam arrives satisfactorily at the pupil of the camera.

17.2 The focus

Stage 1. The grating is rotated on its table until zero order is reflected back to the input slit. A small (5 mm) 90°–45°–45° prism is useful here to shine light from a spectral lamp such as a low-pressure mercury laboratory lamp through the lower half of the input slit. The slit can be opened wide for this adjustment and the prism attached by double-sided sticky-tape. When the grating rotation is correct an image of the slit should appear in the upper half of the slit. A fine adjustment of the grating tilt will allow a Foucault knife-edge test to be made using the slit edge and its image. The eye applied to the slit and looking through it at the M1 mirror will see the illumination appear uniformly on the mirror surface when the focus is correct: otherwise it will appear to wipe across the mirror as the grating rotates. Adjustment is by sliding the slit mechanism in-and-out along the optic axis.

Bear in mind that the focusing cannot be done by turning the M1 mirror to face the entry slit, since the field curvature of the tilted M1 mirror requires the entry slit to be slightly closer to it than its axial focal length (which will have been previously measured) would imply. **For this focus adjustment it is essential that the light be reflected from the grating**.

(This test is very precise and if the operator is unfamiliar with knife-edge testing it can be first practised directly on the M1 mirror without the encumbrance of the input slit mechanism. If the M1 mirror is spherical rather than paraboloid it may show radial zones at the proper focus condition. These correspond to the focus being paraxial or marginal.)

Stage 2. When the M1 focus is satisfactory the grating should be turned to allow a bright first-order line, such as Hg 546.1 nm, coming through the entry slit to fall on to the M2 mirror. A white card is needed here to ensure that the mirror area is fully illuminated. Again the entry slit can be wide open, as only an edge is going to be used. A sharp edge, such as a razor blade, may now be placed at the focus position and the same Foucault test used to adjust the M2 focus position.

If the parts have been made by a good quality workshop all these focus adjustments will be very fine if they are needed at all. The depth-of-focus of a spectrographic instrument working at $\sim F/14$ is $\sim \pm 0.25$ mm and the tolerances on focus are not difficult to achieve. A few millimetres defocus at the entry slit of a 1 m M1 mirror is compensated without significant aberration deterioration by a complementary adjustment of the focal plane of the M2 mirror.

It is the initial alignment of the optic axis which is the sharp and tedious but essential process.

Appendix 1

Optical aberrations

Aberration coefficients

These somewhat complicated expressions are worth including here since they can be set up easily in a computer program and allow the spherical aberration and coma of any lens to be evaluated in a few moments.

Aberration coefficients for lenses

The important quantities are

(a) the spherical aberration coefficient, **B** and
(b) the coma coefficient, **F**.

For a single thin lens with radii of curvature R_1 and R_2 these coefficients are computed from the G-sums:[1]

The G-sums are constructed from eight coefficients which depend only on the refractive index of the glass:

$$G_1 = n^2(n-1)/2$$
$$G_2 = (2n+1)(n-1)/2$$
$$G_3 = (3n+1)(n-1)/2$$
$$G_4 = (n+2)(n-1)/2n$$
$$G_5 = 2(n^2-1)/n$$
$$G_6 = (3n+2)(n-1)/2n$$
$$G_7 = (2n+1)(n-1)/2n$$
$$G_8 = n(n-1)/2$$

and various sums can be computed from these:

The spherical aberration G-sum is

$$G_{\text{sph}} = G_1 c^3 - G_2 c^2 c_1 + G_3 c^2 s + G_4 c c_1^2 - G_5 c c_1 s + G_6 c s^2.$$

---

[1] A. E. Conrady, *Applied Optics and Optical Design* (London: Oxford University Press, 1929), Vol. 1, p. 95.

The coma G-sum is

$$G_{\text{coma}} = G_5 c c_1/4 - G_7 cs - G_8 c^2.$$

Longitudinal spherical aberration $= G_{\text{sph}} \cdot v^2 y^2$
Tangential coma$^2 = -3 G_{\text{coma}} y^2 v \tan \alpha$
where

$v \tan \alpha = h$, the height of the image above the axis
v is the image distance ($= f$, if $s = 0$)
$s, = 1/u$, is the *reciprocal* of the object distance
c_1 is the first surface curvature ($= 1/R_1$)
$c, = \frac{1}{(n-1)} \frac{1}{f} = c_1 - c_2$, is the total curvature of the lens[3]
y is the height of incidence of the marginal ray on the lens.

A special case of particular importance is when the object distance $= -\infty$; that is when rays come from $-\infty$. Then $s = 0$; $v = f$, and equations simplify to

$$G_{0,\text{sph}} = G_1 c^3 - G_2 c^2 c_1 + G_4 c c_1^2,$$
$$G_{0,\text{coma}} = G_5 c c_1/4 - G_8 c^2.$$

Alternatively and conveniently when there is an infinite conjugate on the object side we can define the *aberration coefficients*, **B**, **F₀** and **F** directly from the refractive indices and radii of the lens or mirror by

$$\mathbf{B} = \left(\frac{n}{n-1}\right)^2 \frac{1}{f^2} - \left(\frac{2n+1}{n-1}\right) \frac{1}{R_1 f} + \left(\frac{n+2}{n}\right) \frac{1}{R_1^2},$$

$$\mathbf{F_0} = \left(\frac{n}{n-1}\right) \frac{1}{f} - \left(\frac{n+1}{n}\right) \frac{1}{R_1}.$$

For the lens coefficients there are alternative expressions using the curvatures c_1 and c_2 instead of the radii. They are sometimes useful because they do not involve the focal length:

$$\mathbf{B} = c_1^2[n^3 - 2n^2 + 2]/n - c_1 c_2[2n^2 - 2n - 1] + c_2^2 n^2,$$
$$\mathbf{F_0} = c_1[n^2 - n - 1]/n - n c_2.$$

These expressions simplify greatly if the lens is plano-convex.

Aberration formulae

Longitudinal spherical aberration

$$\Delta_{\text{sph}} = \frac{1}{2} y^2 \mathbf{B} f.$$

The diameter of the 'circle of least confusion'

$$\Delta_{\text{colc}} = \Delta_{\text{blur}} = y^3 \mathbf{B}/4.$$

---

[2] Sagittal coma is 1/3 of this.
[3] Also referred to as Δc elsewhere in the book.

Meridional or tangential coma

The height of the coma patch is given by

$$\Delta_{\text{coma}} = 3y^2(\mathbf{F_0} + \mathbf{ZB})\tan\alpha.$$

Sagittal coma is $1/3$ of this.

Fraunhofer's theorem

The coma coefficient for a thin lens or mirror with its entry pupil at a distance $\mathbf{Z}$ from the vertex is

$$\mathbf{F} = (\mathbf{F_0} + \mathbf{ZB}).$$

Worked example[4]

For a plano-convex lens with a flat first surface ($c_1 = 0$) the second surface curvature is negative (the centre of curvature is to the left of the surface) and the aberration coefficients are

$$\mathbf{B} = c_2^2 n^2, \qquad \mathbf{F_0} = -nc_2,$$

and the lens is free from coma if the pupil is at $\mathbf{Z} = -\mathbf{F_0}/\mathbf{B} = 1/nc_2$, and since c_2 is negative, so is $\mathbf{Z}$ and the pupil is to the left of the lens. In terms of focal length it is at $\mathbf{Z} = -[(n-1)/n]f$.

It is easy to verify that if the first surface of the lens is curved ($c_2 = 0$), there is in general no *real* position for a coma-correcting pupil.[5]

Astigmatism and field curvature

The difference in the curvatures of the two focal surfaces C_s and C_t is given by

$$\Delta C = C_t - C_s = (2/f)(\mathbf{Z}^2\mathbf{B} + 2\mathbf{Z}\mathbf{F_0} + 1).$$

Petzval's theorem

This gives the relation between the Petzval sum $P_z = \Sigma_i \left(\frac{1}{n_i f_i}\right)$, the difference of curvature ΔC and the two curvatures of the sagittal and tangential fields:

$$C_t = \Sigma P_z + (3/2)\Delta C,$$
$$C_s = \Sigma P_z + (1/2)\Delta C.$$

For a mirror, the Petzval sum is found by treating it as a lens with a refractive index -1. Then

$$\Sigma P_z = -1/f.$$

---

[4] From Chapter 4, page 35.
[5] But for a fortuitous exception see Chapter 4, page 37.

Flat-field conditions

Generally a flat field free from astigmatism is obtained only when the coefficients **C** and **D** are both zero. The condition for a flat tangential field in a spectrograph is

$$4\mathbf{C} + 2\mathbf{D} = 0.$$

Appendix 2
Wavelengths of spectral lines for calibration

Laboratory discharge-lamp spectra

Laboratory gas-discharge lamps are commonly available with a wide variety of elemental contents and Table A2.1 contains the wavelengths usually encountered in the popular ones. **Avoid laboratory discharge lamps such as zinc–cadmium–mercury which are operated at high pressure. It is most important that low-pressure lamps be used to ensure narrow spectrum lines which suffer no pressure- or temperature-broadening.**

Table A2.1 lists the wavelengths in angstroms of the bright emission lines in the common low-pressure laboratory discharge lamps. Wavelengths, are *air* wavelengths. Wave*numbers*, for which the unit is cm^{-1}, are always vacuum measurements and must be increased by a factor $(n - 1)$ to convert them to air wavenumbers. The refractive index, n, of air is given to sufficient accuracy for most practical purposes by[1]

$$(n - 1) \times 10^6 = 8340.78 + [2\,405\,640/(130 - \sigma^2)] + [15\,994/(38.9 - \sigma^2)],$$

where σ is the vacuum wavenumber ($= 1/\lambda_{\text{vac}}$).

Solar Fraunhofer lines

These are chiefly for rough calibration of low–medium resolution instruments and the lines are sometimes broad and not well defined. Their advantage is that they are always available, although visual identification is not always easy and an annotated photographic copy of the solar spectrum is useful for field-work. It is worth bearing in mind that although lines are sometimes over 1 Å wide at half-depth the centroids can be located to ±0.1 Å or less, which is adequate for calibration of a spectrographic resolving power of 10 000.

Lines C, D and F are usually used to specify the refractive indices and Abbé v-values of optical glasses and other transparent materials. Some caution is needed. Older astrographic telescope objectives, used with orthochromatic emulsions, were computed to be achromatic in the violet and near UV and used different lines as the standard correction wavelengths.

---

[1] J. C. Owens, *Appl. Opt.*, **6** (1967), 51.

Table A2.1 *Wavelengths (Å) of discharge lamps*

Hydrogen	Mercury	Sodium	Helium	Cadmium	Zinc
6562.79	11287	11404.2	10830	6438.4696	6362.35
4861.33	10140	11382.4	7065.19	5085.82	5181.99
4340.47	5790.65	8194.81	5875.62	4799.91	4810.53
4101.74	5769.59	8183.27	4685.75	4687.15	4722.16
3970.07	5460.75	5895.92	4471.68	3610.51	4680.14
3889.06	4358.35	5889.95	4026.19	3466.20	3345.02
3835.40	4046.56	5688.22	3888.65	3403.65	3282.33
3797.91	3663.28	5682.66		3261.06	
3770.64	3650.15	3303.00		2748.58	
3750.16	2536.52	3302.32			

Notes

(1) The cadmium line at 6438.4696 Å is an old international wavelength standard. Being the result of a singlet transition it is a particularly narrow line and eminently suitable for demonstrating the normal Zeeman effect. It was subsequently replaced by the 6056.125 Å line of krypton-86 but cadmium lamps are still in common use and readily available.

(2) The zinc line at 5181.99 Å is a convenient standard wavelength for measuring small Doppler shifts in a moving source of the Mg triplet, a group of closely spaced Fraunhofer lines in a solar or stellar spectrum.

(3) The HeNe laser line is at 6328.17 Å.

Arc spectra

The spectrum of an iron arc is less convenient than a set of laboratory lamps and the necessary apparatus is mostly to be found in older, well-equipped spectroscopic laboratories. The iron arc in fact is the traditional source for the visible and near UV region, with many hundreds of well-measured wavelengths. Sundry authors have recommended a vertical arc about 12 mm long taking 5 amps from a 230-volt source and with only the central 2 mm of the arc focused on to a horizontal slit. Welder's dark glasses are needed when adjusting it and it is a source of skin-damaging UV.

Alternatively, a hollow-cathode lamp will yield a similar spectrum but at much lower intensity.

Passive wavelength standards

There are sundry methods of calibrating spectrographs which are independent of electricity or sunlight and therefore suitable for field-work at remote sites. Chief among these is absorption spectroscopy using a tungsten lamp, possibly battery-powered, as a light source.

One of the most useful calibrators for the visible region is an absorption cell preferably of glass or quartz, with a diameter of a few centimetres and an optical absorption path length of about 75 cm. It should be evacuated or possibly filled with an inert gas such as helium, and should contain some crystals of iodine. At the equilibrium vapour pressure of

Table A2.2 *Solar Fraunhofer lines*

Letter	Wavelength, Å	Element	Source
A	7621	O	Telluric absorption band
	7594	O	Telluric absorption band
B	6869.95	O	Telluric absorption band
C	6562.79	hydrogen	Balmer Hα
D	5895.92	Na	Sodium resonance line
D	5889.95	Na	Sodium resonance line
E	5269.54	Fe	
b_1	5183.62	Mg	Magnesium triplet
b_2	5172.70	Mg	Magnesium triplet
b_3	5167.51	Fe	
b_4	5167.33	Mg	Magnesium triplet
F	4861.33	H	Balmer Hβ
G'	4340.46	H	Balmer Hγ
G	4307.91	Fe	
	4307.74	Ca	
g	4226.73	Ca	
h	4101.74	H	Balmer Hδ
H	3968.47	Ca II	Ionised calcium resonance
K	3933.67	Ca II	Ionised calcium resonance
L	3820.43	Fe	
M	3727.67	Fe	
N	3581.21	Fe	
O	3440.99	Fe	
P	3361.21	Ti II	
Q	3286.76	Fe	
R	3181.28	Ca II	
	3179.33	Ca II	
s_1	3100.67	Fe	
	3100.30	Fe	
s_2	3099.94	Fe	
T	3021.07	Fe	

iodine at room temperature the electronic transition $X\ ^1\Sigma_g^+ \to B\ O_u^+$ of I_2 gives a well-known series of absorption bands.[2] These are sufficiently deep and have a rotational structure open enough that the narrow individual absorption lines give a large number of accurate calibration points throughout the yellow and green. The amount of absorption and therefore the depth of the lines depends markedly on the ambient temperature but the wavelength calibration is very precise.

More compact, with some absorption lines in the visible but a large number in the UV, is europium.[3] An absorption cell with an absorption path length a few millimetres long

[2] I. J. McNaught, *J. Chem. Educ.*, **57** (1980), 101.
[3] G. Smith & F. S. Tomkins, *Proc. Roy. Soc.*, **A342** (1975), 149; *Phil. Trans. Roy. Soc.*, **A283** (1976), 345.

filled with an aqueous solution of europium chloride is highly portable. The lines result from atomic transitions betweeen energy levels of unfilled inner electron shells in the europium atom. The lines are not generally as narrow as those in the vapour phase but as calibration marks are comparable with solar Fraunhofer lines.

Appendix 3

The evolution of a Fabry–Perot interference spectrograph

This is to show the stages of evolution of a design for a Fabry–Perot spectrograph for night airglow or astronomical applications.

(1) Astronomical considerations required that the resolution should be 1.2 Å at 5183.6 Å, the wavelength of the Mg triplet of solar Fraunhofer lines. The free spectral range was to be about 20 Å, set by the FWHM of the interference filter which was to act as order-sorter. The available finesse was 40 and this constrained the étalon gap thickness to be 28 μm. The order at the centre of the ring system is then 56 μm/0.5183.6 μm = 108.

(2) The available CCD camera used a 50 mm F/1.4 'Takumar' lens and a CCD with 1152×325 pixels each 25 μm square. The field semi-angle was then $16°$. A primary requirement was for spatial resolution, so that both the sky and the Fabry–Perot fringes were to be in focus on the CCD. It was also required that the angle subtended by the CCD on the sky should not be greater than $\sim 5°$.

(3) The étalon gap thickness determined the angular radii of the Fabry–Perot rings (assuming $m_0 = 108$ at the centre) which are given by $\cos \theta_n = (m_0 - n)/m_0$ with $n = 1, 2, 3 \ldots$ so that $\cos \theta_1 = 7.8°$. The next rings are at $11.04°$ and $13.53°$. Their diameters on the CCD are at $50 \text{ mm} \tan \theta_n$ which are 0, 13.7 mm, 19.5 mm and 24.07 mm, so that the CCD records approximately three orders of interference. This would be so even if the order at the centre is not integer.

(4) Conservation of étendue then required a telescope with an aperture[1] 32/5 times that of the camera lens, i.e. 160 mm. The design then compromised with an available plano-convex lens of 6″ (152 mm) aperture and 1000 mm focal length. This would give adequate resolution on the sky and the small spectral range meant that chromatic aberration was no problem. It remained to collimate the image so that the sky could be received at the CCD with the Fabry–Perot fringes superimposed on it. At the same time this collimator had to image the objective on to the pupil of the camera. The result was the configuration shown in Fig. A3.1a in which the image collimator needed is clearly inconveniently large. This diagram, like the others, is not to scale, the vertical dimension having been greatly exaggerated for clarity.

(5) The field of the objective required by the camera is in the ratio (again conserving étendue) of their pupil diameters, so that the field semi-angle for the objective is $25/150 \times 16° = 2.66°$. The actual field radius is then 46.5 mm, requiring the order-sorter to have an effective diameter of 93 mm. In the actual instrument a standard 100 mm diameter interference filter was used.

---

[1] The ratio of the field angles.

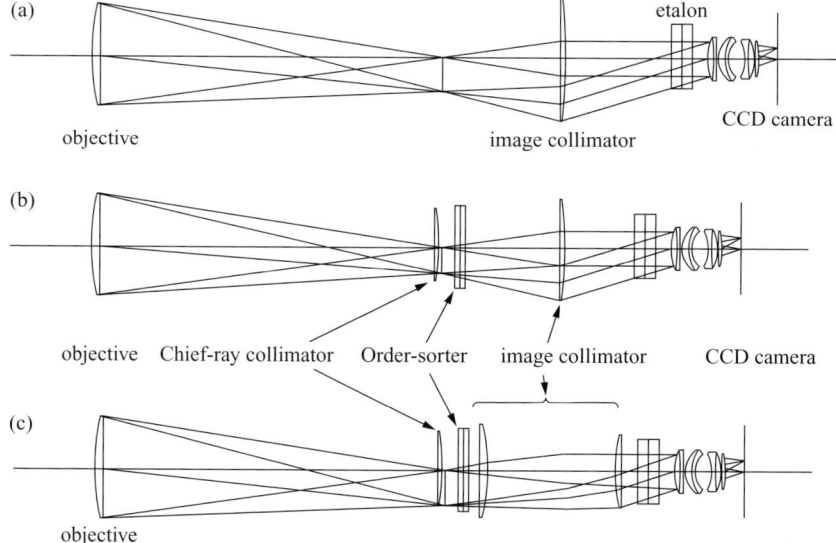

Figure A3.1 The evolution of a Fabry–Perot CCD spectrograph following the precepts of Chapter 10. The vertical scale has been exaggerated for clarity. In practice a 150 mm diameter objective would result in an instrument 2 m long.

(a) The primitive design. Objective – collimator – camera.

(b) The addition of a chief-ray collimator immediately to the left of the order-sorting interference filter moves the exit pupil (and hence the CCD camera) in to the F_1 focal plane of the image collimator.

(c) The splitting of the image collimator into two smaller lenses, with the same focal length and P_1 principal plane as the original singlet lens, reduces the overall diameter of the spectrograph to that of the objective. Ideally the first member of the pair should have a focal length which makes the marginal ray parallel to the optic axis, but in practice one may compromise with the contents of a manufacturer's catalogue.

(6) It was a sine qua non that the order-sorter should have all chief-rays parallel to the optic axis as they passed through it. This was achieved by inserting, just in front of the focus of the 1000 mm objective lens, a chief-ray collimator lens of sufficient diameter (100 mm) to cover the necessary 5° field and of 1000 mm focal length. This came from the same catalogue as the objective.

The order-sorter could then be placed close to, but not exactly at the focus of the objective. Any other position for the order-sorter would have entailed steeper marginal rays than those from the F/6.66 of the objective and would have reduced the efficiency of the component.

(7) This chief-ray collimator made no difference to the image collimator but meant that its diameter could be reduced somewhat. The other effect was to move the exit pupil of the system to the back-focus of the image collimator, shortening the system slightly.

These various criteria resulted in the configuration shown in Fig. A3.1b. A scale drawing of the optical layout then determined the diameter needed for the étalon plates.

(8) The image collimator was still inconveniently large in diameter and so, with the object of reducing the diameter of the whole instrument, it was replaced by a pair of lenses with the same focal length and the same positions for the principal planes. There was some latitude in design here and items from the catalogue were found which would give the result required.

This resulted in the final form of the instrument shown in Fig. A3.1c.

Appendix 4

The common calibration curve in silver halide spectrophotometry

The problem of photometry with silver halide emulsions is that of finding the relation between density and exposure. Density is defined as minus the logarithm of the transmission ratio, i.e. of the fraction of incident light transmitted through the emulsion to the detector (on the assumption that the response of the detector is linear). Exposure is defined as the product of incident intensity during the photographic exposure and the duration of the exposure, and it is assumed, justifiably, that the density depends only on the product.

The method of the common calibration curve[1] is to take several photographs of the same spectrum under identical circumstances, but with different neutral-density filters, the filter densities ascending for example in factors of 2. The densities of a particular spectrum line, measured from the different exposures, will give a set of points on the calibration curve.[2] A different spectrum line can then be interpolated on this initial line and the different exposures of this second line provide further points which extrapolate the curve at one end or the other. By continuing in this fashion with other spectrum lines, the curve can be continued, always by *interpolating* new initial material, gradually extending the curve in both directions as far as necessary and well beyond the linear portion.

The result is not of course absolute, but will allow comparison of intensities (provided the wavelength range is not too great) with an accuracy of up to $\pm 1\%$ provided that proper care has been taken with all the processing details.

---

[1] I have been unable to find any reference to the original inventor of this technique. I myself learned it and used it while doing spectrography in R. W. Ditchburn's XUV laboratory in Reading University during the last century.
[2] Two will do at a pinch. Three or four is better.

Bibliography

Blackman, R. B. & Tukey, J. W., *The Measurement of Power Spectra* (New York: Dover, 1959).
Braddick, H. J. J., *The Physics of Experimental Method* (London: Chapman & Hall, 1963).
Conrady, A. E., *Applied Optics and Optical Design* (London: Oxford University Press, 1929).
Harrison, G. R., Lord, R. C. & Loofbourow, J. R., *Practical Spectroscopy* (New York: Prentice-Hall, 1948).
James, J. F. & Sternberg, R. S., *The Design of Optical Spectrometers* (London: Chapman & Hall, 1969).
Kingslake, R., *Lens Design Fundamentals* (New York: Academic Press, 1978).
Kingslake, R., *A History of the Photographic Lens* (San Diego, CA: Academic Press, 1989).
Loewen, E. G. & Popov, E., *Diffraction Gratings and Applications* (London: Marcel Dekker Ltd., 1977).
Moore, J., Davis, C., Coplan, M. & Greer, S., *Building Scientific Apparatus* (Boulder, CO: Westview Press Inc., 2005).
Palmer, C., *Diffraction Grating Handbook*, 6th edn. (Rochester, NY: Newport Corporation, 2005).
Rieke, G., *Detection of Light* (Cambridge University Press, 2002).
Stroke, G. W., 'Diffraction Gratings', in *Handbook of Physics*, Vol. 29 (Berlin: Springer, 1967).

Index

aberrations, 28
aberration coefficients
  chromatic, 39
  for lenses, 175
  for mirrors, 37
  formulæ, 176
  of Rowland grating, 97
aberration theory, Seidel, 29
Abney mounting for concace grating, 93
absorption
  cells, 136
  lines, 182
  measurements, 170
addition theorem, 42
Airy function, 102
aliasing, 72
alignment, optical, 142, 156, 172
alignment tool, 173
aperture, 20
  numerical, 20
aplanatic lens, 16
aplanatic surface, 15
apodising, 68
  masks, 69
arc spectra, 181
astigmatism, 32, 33, 177
autocorrelation function, 45, 115
auxiliary optics, 128

baffles, 81, 141
  external, 135
bearings, journal, 166
  angular contact, 158
Bessel functions, 71
Beutler eikonal theory, 97
black body radiation, 25, 26
blazing, of gratings, 67
Bunsen, R., 1

Calibration
  CCD sensitivity, 126
  lines, laboratory, 180
  lines, solar, 181
  passive wavelength standards, 180
  spectrograph, 126
  wavelengths, 169, 180
cameras, CCD, 124
camera lenses, 40, 60, 85, 132
  thorium in, 125
Cassegrain focus, 132
centred systems, 28
charge-coupled device (CCD), 123 *et seq*.
circle of least confusion, 30
circular slit theorem, 75
cleaning of optical materials, 164
  diffraction gratings, 164
coma
  control with a stop, 35, 36
  equations for, 31, 35, 176,177
  sagittal, 31, 82, 176
  Seidel coefficients, 38
  tangential, 81, 145, 177
coma patch, height of, 31, 177
common calibration curve, 185
configurations, spectrograph
  Abney, 93, 94
  alternative, 84, 85
  Čzerny–Turner, 80
  Eagle, 94, 95, 99
  Ebert–Fastie, 77
  Littrow prism, 60
    grating, 78
  Paschen–Runge, 93
  Pellin–Broca, 60, 61
  Pfundt, 79
  Rowland, 93
  Seya–Namioka, 96, 97
  Johnson–Onaka, 96
  Wadsworth stigmatic, 94, 95
Connes, P., 4
Conrady, A. E., 28, 111, 133, 175
convolutions, 42
convolution algebra, 44
convolution theorem, 44
cosmic ray events (CREs), 124
Coudé focus, 41, 53

Index

Courtès, G., 132
Crookes, Sir William, 2
cubic splines, 169
curvature, astigmatic, 97
curvature, field, 33, 177
curvature, surface, *see* surface curvature
Čzerny–Turner mounting, 77, 80 *et seq.*
  aberrations of, 81 *et seq.*
  alignment of, 172
  asymmetrical, 158
  crossed, 81
  vertical table, 153
  worked example of, 145

decentred stop, 84
dispersion, 76
Diffraction, Fraunhofer, 52
  single slit, 53
  two-slit, 54
  N-slit, 54
  with oblique incidence, 54
Dirac comb, 48
Dirac delta function, 48
dispersion relation, in CCD spectrographs, 123
dispersion
  and resolution, 4
  curve for quartz, 59
  differential, 76
  partial, 6, 39
distortion, 34
doublet, achromatic, 39
DQE (detector quantum efficiency), 3

Eagle mounting for concave grating, 94
Ebert, H., 2
Ebert–Fastie mounting, 77
echelle, 3
Edser–Butler fringes, 66
emulsions, silver halide, 5, 76, 121
equivalent width, 170
étalon, *see* Fabry–Perot
étendue, 5, 24, 62

F-number, 20
Fabry, C., 3
Fabry lens, 129, 130
Fabry–Perot
  étalon, 102
  monochromator, 104
  spectrometer, 105
    mechanical scanning, 105
    CCD spectrograph, 108, 183
    spectrum extraction, 111
  theory, 102
Fellgett, P., 5, 117
fibre optical input, 137
field, 21
  angle, 31, 34, 81
  stop, 21
flat-field condition, 83
focal isolation, 61

focal reducer, 132
  in Čzerny–Turner spectrograph, 133
focus, as zero-order aberration
  paraxial, 31
  sagittal, 32
  tangential, 32
fore-optics, 128
Fourier's inversion theorem, 41
Fourier spectrometry, principles, 114
Fourier transforms in optics, 52
Fraunhofer, J. von, 1
Fraunhofer, diffraction, *see* diffraction, Fraunhofer
Fraunhofer, lines, 180, 181
free spectral range, 69, 103
FWHM (full width at half maximum), 46

Gauss, C. F., 12
Gaussian function, 46, 112, 137
Gaussian optics, 12, 28
ghosts
  Lyman, 71
  Rowland, 70
glass, 39, 58, 163
  wavelength dispersion, 59
grating, diffraction, 63 *et seq.*
  cleaning, 164
  concave, 89 *et seq.*
  holders, 156
  transmission, 85
  turntable, 158

Helmholtz–Lagrange invariant, 23
Herschel, Sir William, 1
Huygens, J. C., 52

incidence, oblique, 54
infra-red spectrometry, 62, 114
  emulsions for, 121
iris diaphragm, 20 *et seq.*, 84

Jacquinot, P., 4, 105
Johnson–Onaka mounting, 96

Kinematic mounting, 151
  principles, 166
Kingslake, R., 40, 145
Kirchhoff, G., 1

Lambert's law, 25
  radiator, 126
lenses
  coma of, 36, 175
  double-Gauss, 60, 134
  focal planes and points, 14, 18
  holders, 153
  nodal planes and points, 19
  plano-convex, 36, 132, 145, 176
  principal planes and points, 18
  Tessar, 60
  thick, 15
lensmaker's equation, 15

Littrow mounting
  grating, 78
  prism, 59
LN$_2$, 120, 161
Lorentz function, 46
luminosity, 24, 26, 112
luminosity–resolution product, 4
Lummer, O., 3

mechanical design, 150
Meinel, A. B., 132
meridional coma, *see* tangential coma
Mertz, L., 2, 4
Michelson, A. A., 2
Michelson Fourier spectrometers, 62, 114 *et seq.*
  cat's eye, 116
  cube-corner, 116
  tilting, 116
mirrors
  adjustments, 155
  holders, 153
  paraboloidal, 30, 36, 38, 77, 80, 163
  spherical, 30, 38, 76, 163
multiplex spectrometer, 114
  advantage, 117
multiplication theorem, 42
multislit scanning, 108

Newtonian mirrors, 79, 111
numerical aperture, 20

oblique incidence, 54
Onaka, R., 96
optical design, 139
  layout, 139, 150
  transparent materials, 163
  refinement, 29, 140 *et seq.*
optical screw, 106
order overlap, 68

paint
  optical black, 167
  protective, 167
Parseval's theorem, 42
Paschen–Runge mounting for concave grating, 93
path
  geometrical, 11
  optical, 11
  reduced, 12
Pellin–Broca prism, 60
Peltier cooling, 123
Periodic errors, 70
Perot, A., 3
Petzval sum, 37
Petzval's theorem, 37
Pfundt mounting, 79
phase angle, 101
photography, silver halide, 3, 76, 92, 120, 185
photomultipliers, 122
piezo-electric transducers, 105, 106
pixels, 84, 108, 123, 145, 169, 183

Planck formula, 26
pressure scanning, 107
prism spectrographs, 57
  Pellin–Broca, 61
  Littrow, 60
Profile, instrumental, 43, 47, 69, 103
  Fabry–Perot étalon, 111
pupil, control of aberrations by, 21
pupils, laterally displaced, 84

'Q' plates, for XUV spectrography, 122
quartz, wavelength dispersion of fused, 59

Ramsden, J., 132, 145
ray-tracing, computer, 140
  methods, 142
  programs, 144
Rayleigh criterion, 58
reciprocity failure, 3, 5, 120
refractive index, 11, 58, 85
  determination of, 57
  of mirror, 15
resolution, 58
resolving power
  defined, 24
  of diffraction grating, 64
  of Fabry–Perot étalon, 105
  of prism, 58
Rosendahl condition, 81
Rowland, H. A., 2
Rowland circle, 89
Rowland grating theory, 89 *et seq.*
Rowland mounting for concave grating, 93
Rubens, H., 61

sagittal coma, 31
sagittal focus, 33
sampling theorem, 49
scattering
  causes of, 135
  from edges, 135
Schmidt camera spectrography, 134
Schuster, Sir Arthur, 2
screw, optical, 106
seidel aberrations of decentred stop, 84
Seya–Namioka mounting, 96
shift theorem, 42
shroud, spectrograph cover, 167
sinc function, 46
slitless spectrography, 131
slits, construction of, 159, 161
spectral power density (SPD), 45, 115
spectro, slitless, 131
spectrograph mountings, *see* configurations
spectrum lines, 179, 180, 181
  shapes of, 47
spherical aberration
  Seidel coefficient, 30
  equation for, 176
  indicator, 145
spot diagrams, 148

Stefan's constant, 25
stops, 20, 28
  field
    lateral displacement of, 129
    telecentric, 84
sulphones, 163
surface brightness, 25
surfaces curvature
  focal, curvature of field, 33 *et seq.*, 177
  focal, flat field condition, 83
  Petzval, 37
  sagittal, 37
  tangential, 37

tangential coma, 85, 145, 177
tangential focus, 94, 96
telecentric operation, 129
  stops, 129
telescope
  astronomical, 130
  refracting, 22
temperatures
  addition, 42
  circular slit, 75
  convolution, 44
  focal curve, 59
  Fraunhofer's, 177
  Parseval's, 42
  Petzval's, 177
  sampling, 49
  shift, 42, 45, 67
  Wiener–Khinchine, 45
top-hat function, 45

ultra-violet, 7, 122, 163, 181
UVOIR, 7

vacuum ultra-violet, 8
vertex, of optical surface, 12
vignette, 22
Voigt profile, 47

Wadsworth stigmatic mounting, 92, 94
wavelength calibration, 126, 169, 179
  standards, 168
    passive, 180
    tables, 180 181
White cells, 136
Wiener–Khinchine theorem, 45
Wollaston, W. H., 1
Wood, R. W., 1

XUV, 7, 71, 92, 96
  photographic plates for, 122

zero meridional coma (ZMC), 140